Algebraic Probabilistic Consistency

Volume 69

ALGEBRAIC PROBABILISTIC CONSISTENCY
Studia Logica Library

VOLUME 69

The book series Trends in Logic covers essentially the same areas as the journal Studia Logica, that is, contemporary formal logic and its applications and relations to other disciplines. The series aims at publishing monographs and thematically coherent volumes dealing with important developments in logic and presenting significant contributions to logical research.

Volumes of Trends in Logic may range from highly focused studies to presentations that make a subject accessible to a broader scientific community or offer new perspectives for research. The series is open to contributions devoted to topics ranging from algebraic logic, model theory, proof theory, philosophical logic, non-classical logic, and logic in computer science to mathematical linguistics and formal epistemology. This thematic spectrum is also reflected in the editorial board of Trends in Logic. Volumes may be devoted to specific logical systems, particular methods and techniques, fundamental concepts, challenging open problems, different approaches to logical consequence, combinations of logics, classes of algebras or other structures, or interconnections between various logic-related domains.

This book series is indexed in SCOPUS.
Authors interested in proposing a completed book or a manuscript in progress or in conception can contact either chris.wilby@springer.com or one of the Editors of the Series.

Daniele Mundici

Algebraic Probabilistic Consistency

Boole, Łukasiewicz, de Finetti, Kolmogorov

Springer

Daniele Mundici
Department of Mathematics and Computer
Science
University of Florence
Florence, Italy

ISSN 1572-6126 ISSN 2212-7313 (electronic)
Trends in Logic
ISBN 978-3-031-98336-8 ISBN 978-3-031-98337-5 (eBook)
https://doi.org/10.1007/978-3-031-98337-5

This Springer imprint is published by the registered company Springer Nature Switzerland AG
The registered company address is: Gewerbestrasse 11, 6330 Cham, Switzerland

1991 Mathematics Subject Classification. Primary: 03B50, 60A05. Secondary: 03B05, 03B25, 03G05, 03G20, 06D35, 06E05, 06F20, 28A35, 28A60, 52B11, 52B12, 52B55, 60A10, 60B05, 68T37.

Key words and phrases. Boolean algebraic states; de Finetti's consistency theorem; Dutch Book theorem; Boole's probabilistic inference problem on the object of probability; de Finetti's "Fundamental Theorem of Probability"; probabilistic consistency and PSAT; the product law for independent Boolean algebras; independence and the product law; characterization of Łukasiewicz logic; MV-algebra; the MV-algebraic probabilistic consistency theorem; Boole's probabilistic inference problem for continuous events; independence and the product law for continuous events; finite vs. countable additivity; MV-algebras with a state and Kolmogorov triples on compact Hausdorff spaces; the conditional $PSAT_{\infty}$ problem; syntactic and semantic consequence in Łukasiewicz logic; decidability of conditional probabilistic consistency; product states; exchangeability theorem; de Finetti theorem.

To my students in Africa, South America, and Europe

To Lihuén

Competing Interests The author has no competing interests to declare that are relevant to the content of this manuscript.

Introduction

This book is primarily aimed at undergraduates in mathematics and computer science who wish to understand the symbiosis/equivalence between de Finetti's finitely additive probability and Kolmogorov's countably additive probability for continuous real-valued random variables on compact Hausdorff spaces. The book may also be of interest to philosophy students interested in the mutual relationship between consistency in logic and in probability, and to researchers interested in the algebraic approach to probability.

The *attraction force* that determines the symbiosis between finitely additive and countably additive probability is given by de Finetti's definition of a consistent $[0, 1]$-valued map on a finite set of $\{0, 1\}$-valued events. This is a generalization of a logically consistent assignment of truth-values in $\{0, 1\}$ to a finite set of premises.

The *main application* of this equivalence is the decidability of the consistency of a rational-valued assignment of probabilities to a finite set of continuous $[0, 1]$-valued events conditioned by another finite set of events. This is made possible by the deductive-algorithmic machinery of Łukasiewicz infinite-valued logic $Ł_\infty$, whose formulas code $[0, 1]$-valued events just as boolean formulas code $\{0, 1\}$-valued events.

The *key role* of $Ł_\infty$-formulas is a consequence of the fundamental fact that $Ł_\infty$ is the unique logic generated by a continuous $[0, 1]$-valued operation satisfying the two minimalist conditions of what is commonly understood as an implication.

Last, but not least, a *notable feature* of the equivalence between finitely additive and countably additive probability proved in this book is that logic and algebra play first violin in the symphony of mathematical results culminating in the proof of the equivalence. All these results will be the described in the final part of this introduction.

Boole, de Finetti, Łukasiewicz, Kolmogorov. Every day we make probabilistic estimates about tomorrow's weather conditions, gasoline prices, traffic intensity, the punctuality of trains and friends, etc. And from our probability assessments we try to guess the probability of other events related to them.

On page 246 of his 1854 book "Investigation of the Laws of Thought", Boole writes that the object of probability theory can be defined as follows: Given the probabilities of any events, of whatever kind, to find the probability of some other event related to them. This is known as *Boole's probabilistic inference problem.*

To fix ideas, let us suppose we are given the probabilities of the following events:

Event with given probability	Event Name	Its probability
Gold will reach \$6000 an ounce by 2026	E1	0.01
The next pope will be African	E2	0.4
The next pope will be Asian	E3	0.15

Next let us consider some probability assignments for other related events:

Related event	Probability	Probability law violated by the assignment
E2 or E3	0.7	Additivity for disjunctions of incompatible events
E1 and E2	0.3	Product law for logically independent events
E2 or not E2	0.9	The probability of the sure event must be 1
E1 and E2	−0.6	Probability values must lie in the real interval [0, 1]

Where do these probability laws come from? How do they affect the probabilities of other events related to them? Why cannot we assign negative probabilities, or probabilities greater than 1? To answer these questions, in his 1931 paper published in Fundamenta Mathematicae, de Finetti introduced an operational notion called by him "coerenza" ("*consistency*" in English) and showed that all laws of probability follow from his definition of a "consistent" [0, 1]-valued assignment to finitely many events. Here is the English translation of what he writes on page 328 of his paper, in his own italics:

> Having proved the fundamental properties of the classical probability calculus, it follows that all its results are nothing else but *consequences* of the definition of *consistency* given in this paper.[1]

Having little inclination for mathematical logic, de Finetti didn't notice that his notion of "coerenza" boils down to logic consistency in the particular case when the values assigned to the events restrict to $\{0, 1\} = \{\text{no, yes}\} = \{\text{false, true}\}$. Indeed, paraphrasing Boole, a main "object of boolean logic", understood as the analysis of boolean consequence, contradiction, independence of a new formula given a set of assumptions, may be thus defined: Given the truth-values 0 (false) or 1 (true) of any assumptions, of whatever kind, to find the possible truth-values (if any) of some other proposition connected with them.

De Finetti's consistency notion can be introduced with the help of an *ideal* bookmaker who posts on the internet betting odds $p_1, \ldots, p_h$ on the possible outcomes

[1] Dimostrate le proprietà fondamentali del calcolo classico delle probabilità, ne scende che tutti i risultati di tale calcolo non sono che *conseguenze* della definizione che abbiamo data della *coerenza*.

of events $E_1, \ldots, E_h$, such as football matches or horse races. Bettors can make positive, as well as *negative* bets on these events. A negative bet s_i on event E_i swaps the bookmaker/bettor roles as follows: The bookmaker now pays $p_i \cdot |s_i|$ to the bettor, in the hope of winning $|s_i|$ if the event E_i occurs. (Readers may choose their preferred currency for all financial transactions in this game.) Following de Finetti, the set $p_1, \ldots, p_h$ of bookmaker's betting odds on the events $E_1, \ldots, E_h$ is "inconsistent" if a smart bettor can place bets $s_1, \ldots, s_h$ on these events that ruin the bookmaker—*whatever the outcome of the events.*

De Finetti himself pointed out that if any of the odds p_i offered by the bookmaker does not lie in the unit real interval, a bettor can easily bankrupt him by a smart choice of the bets $s_1, \ldots, s_h$.

Needless to say, many sets of betting odds p_i posted on the internet are inconsistent—but those who post these odds rarely go bankrupt, because bettors are not allowed to place negative bets.[2] To make their odds more attractive, bookmakers publish their "odds" as $1/p_i \geq 1$ rather than $p_i \in [0, 1]$.

The attentive reader may have noticed that the events $E_1, \ldots, E_h$ and their mutual relations have been introduced so far as primitive notions, like "point" , "space", or "quantity". The same goes for the expression "whatever the outcome of the events". Precise definitions of all these notions are given in the following table:

Primitive Notion	Mathematical Definition
Events $E_1, \ldots, E_h$	Members of a boolean algebra A
The opposite of event E	The complement $\neg E$ of E in A
E or E'	The disjunction $E \vee E'$ in A
E and E'	The conjunction $E \wedge E'$ in A
For any outcome of the events	For any homomorphism of A into $\{0, 1\}$

The author is confident that this table has not left the reader baffled and lost.[3] Using the precise language of mathematics a deep theory will emerge in this book, where two seemingly opposing approaches to probability are shown to be equivalent.

The text is divided into two parts: Part I presents de Finetti's consistency notion for boolean yes-no events and its main applications. All necessary preliminary material on boolean algebras, their homomorphisms and their states (= finitely consistent maps) is presented systematically and with complete proofs. The first three chapters and a large portion of the fourth are self-contained. The rest of Part I needs Stone duality in combination with Carathéodory's extension theorem from a state σ of a boolean algebra A to a probability measure P_σ on the Stone space Ω_A of A. Part I culminates with the result stating that the map $A \mapsto \Omega$, $\sigma \mapsto P$ yields a one-one

[2] Some readers will find our ideal bookmaking somewhat reminiscent of what today is known as "asset pricing without discount".

[3] As Goethe used to say: Mathematicians are like Frenchmen: whatever you say to them they translate into their own language and forthwith it is something entirely different.

correspondence between boolean algebras A equipped with a finitely consistent map σ, and the totality of Kolmogorov triples $(\Omega, \mathcal{B}(\Omega), P)$, where P is a probability measure on the space $\mathcal{B}(\Omega)$ of Borel sets of the boolean space Ω.

In Part II de Finetti's consistency notion is extended to continuous $[0, 1]$-valued events. All results in this part of the book are generalized from boolean logic and boolean algebras to Łukasiewicz infinite-valued logic $Ł_\infty$ and their algebras, known as Chang MV-algebras. The choice of $Ł_\infty$ and MV-algebras among many possible candidates is determined by a recent characterization theorem, whose proof is the subject matter of the first chapter of Part II. Then the above transformation $A \mapsto \Omega$, $\sigma \mapsto P$ is extended to MV-algebras and their states, and is shown to yield all Kolmogorov triples on compact Hausdorff spaces. (Throughout this book, the terms "map", "function", "transformation" will be used interchangeably.)

The rest of this introduction is devoted to a detailed description of the main mathematical results of the book.

Content of Part I, Chapters. 1–4. To begin with a fundamental notion that will pervade this book, let us agree to say that a *state* of a boolean algebra A is a map $\sigma : A \to [0, 1]$ which is additive on disjunctions of incompatible elements and satisfies the (normalization) condition $\sigma(1) = 1$. Notable examples of states of A are given by the elements of the set $\hom(A)$ of homomorphisms of A onto the two-element boolean algebra $\{0, 1\}$. States are also known as "finitely additive probability measures".

For any finite subset $E = \{h_1, \ldots, h_m\}$ of a boolean algebra A and $\{0, 1\}$-valued map β defined on E let us consider the following condition:

For all $s_1, \ldots, s_m \in \mathbb{R}$, $\sum_{j=1}^{m} s_j\big(\beta(h_j) - \eta(h_j)\big) \geq 0$ for some $\eta \in \hom(A)$. (C)

We then have: The map β satisfies condition (C) if and only if it can be extended to a homomorphism of A into $\{0, 1\}$, i.e., β is *logically consistent.*

Indeed, if β is extendable to some $\eta \in \hom(A)$ condition (C) trivially holds. If β is not extendable to any $\eta \in \hom(A)$ then let $s_j = 1$ for all j with $\beta(h_j) = 0$, and $s_j = -1$ for all j with $\beta(h_j) = 1$. Then $s_1, \ldots, s_m \in R$, $\sum_{j=1}^{m} s_j\big(\beta(h_j) - \eta(h_j)\big) \leq -1$ for any $\eta \in \hom(A)$, and condition (C) fails. This completes the proof.

Generalizing this result from $\{0, 1\}$-valued to $[0, 1]$-valued maps defined on E, in Theorem 3.8 we will prove de Finetti's consistency theorem:[4]

> For any boolean algebra A and set $E = \{h_1, \ldots, h_m\} \subseteq A$, a map $\beta : E \to [0, 1]$ satisfies condition (C) if and only if β is extendable to a state of A.

Whenever a function $\beta : E \to [0, 1]$ satisfies condition (C) we naturally say that β is *consistent*. The same terminology was adopted by de Finetti in his original proof of this theorem.[5] Theorem 3.8 is the main tool to establish the close relationship between consistency, independence, and the countable additivity axiom in

[4] also known as his "Dutch Book" theorem.

[5] The Italian adjective "coerente" adopted by de Finetti in his original paper is translated "consistent" in the present book for the sake of terminological consistency. As a matter of fact, the Italian

Kolmogorov's probability theory. The self-contained proof of the theorem given in this book should be of help to readers interested in algebraic probabilistic consistency but not conversant with all the multifarious ingredients of the proof.

As a preliminary result, in Theorem 1.12 a simple proof is given of what de Finetti called "the Fundamental Theorem of Probability". With reference to Boole's probabilistic entailment problem mentioned at the outset of this introduction, the theorem states that the possible probabilities that can be assigned to a new event given the probabilities of old events, form a closed subinterval of the real unit interval—which is empty precisely when the assignment is inconsistent.

In Chap. 4 the classical notion of boolean algebraic "independence" (Definition 4.1) is combined with Theorem 3.8 to obtain the boolean algebraic "product law". When we say, e.g., that Brazil's victory in the next football World Cup is "independent" of the discovery of a new massless particle next year, we are using the logic-algebraic notion of independence, where no probability estimates are assigned to events. Theorem 4.3 establishes the following "product law":

> Product is the only continuous $[0, 1]$-valued operation $\diamond$ on $[0, 1]^2$ which preserves the consistency of the probabilities (betting odds) of conjunctions of $\{0, 1\}$-events in any two logically independent consistent sets.

The proof of the theorem stating that at most one consistency preserving operation $\diamond$ exists is elementary, like all proofs preceding that theorem in the book. On the other hand, to prove that product does preserve the consistency of conjunctions of events in logically independent sets, we will need Stone duality (4.6), Carathéodory's extension theorem (4.10), and the product measure theorem (4.12). For the proofs of these three results detailed references will be provided to the reader in the final section of Chap. 4.

Building on Theorem 4.3, Theorem 4.14 concludes Part I of the book with the following result:

> Boolean algebras A with a state σ are in one-one correspondence with Kolmogorov triples $(\Omega, \mathcal{B}(\Omega), P)$, where Ω is any arbitrary boolean space, $\mathcal{B}(\Omega)$ is the space of Borel sets of Ω, and P is any regular probability measure on $\mathcal{B}(\Omega)$. The correspondence maps A to the Stone space Ω of A and transforms σ into the Carathéodory extension P of σ to $\mathcal{B}(\Omega)$.

For short, de Finetti finitely additive probability for yes-no events is equivalent to Kolmogorov countably additive probability for continuous events on boolean spaces.

Content of Part II, Chaps. 5–10. In the second part of the book we will refine the notions introduced in the first part for yes-no events, and extend Theorem 4.14 to all Kolmogorov triples $(\Omega, \mathcal{B}(\Omega), P)$ *for continuous real-valued events on any compact Hausdorff space* Ω.

---

translation of "logical consistency" is "coerenza logica", and—as we have just seen—de Finetti's consistency is a straightforward generalization of logical consistency. Also see Corollary 3.14. Let us finally note that G. Majone, the translator of most papers included in de Finetti's 1972 book "Probability, Induction and Statistics", Wiley, London, 1972, uses "consistency" instead of "coherence".

In his 1933 book on probability Kolmogorov introduces the countable additivity axiom for his probability triples $(\Omega, \mathcal{B}(\Omega), P)$ as follows:

> We limit ourselves, arbitrarily, to only those models which satisfy Axiom VI. This limitation has been found expedient in researches of the most diverse sort.
>
> A.N. Kolmogorov, "Grundbegriffe der Wahrscheinlichkeitsrechnung". Springer, Berlin, 1933. English translation in: "The Foundations of the Theory of Probability", Chelsea Publishing Company, New York, 1950. See p.15 in the English translation.

His Axiom VI is equivalent to countable additivity. In Kolmogorov probability, "events" are understood as elements of the σ-field $\mathcal{B}(\Omega)$ of Borel subsets of the set Ω of "elementary events", and P is a probability measure on $\mathcal{B}(\Omega)$.

To extend the scope of algebraic probabilistic consistency to Kolmogorov probability on compact Hausdorff spaces, we need a generalization of boolean logic to a logic whose formulas code [0,1]-valued events on compact Hausdorff spaces, understood as (normalized) continuous random variables on these spaces. The self-contained proof of Theorem 5.6 yields the unique solution to this problem:

> Łukasiewicz logic $Ł_\infty$ is the only logic generated by a $[0, 1]$-valued continuous operation $\rightarrow$ defined on $[0, 1]^2$ satisfying the two minimum conditions of what is commonly understood as an "implication", *i.e.*, $[x \rightarrow y = 1$ iff $x \leq y]$ and $[x \rightarrow (y \rightarrow z) = y \rightarrow (x \rightarrow z)]$ for all $x, y, z \in [0, 1]$.

Introduced by C.C. Chang in 1958, MV-algebras are the algebras of Łukasiewicz logic. Boolean algebras are a subclass of MV-algebras. The theory of algebraic probabilistic consistency for continuous $[0, 1]$-valued events rests upon the following MV-algebraic extension (Theorem 6.3) of the boolean consistency Theorem 3.8:

> A $[0, 1]$-valued function β defined on a finite subset E of an MV-algebra A is consistent (as per condition (C)) if and only if it can be extended to a state of A.

Here a *state* σ of an MV-algebra A is a $[0, 1]$-valued function on A which is additive on incompatible elements and satisfies the normalization condition $\sigma(1) = 1$. By Corollary 6.14, a $[0, 1]$-valued function f defined on an MV-algebra A is a state if and only if every finite restriction β of f satisfies the consistency condition (C). For short, f is *finitely consistent.* The generalization of the "Fundamental Theorem of Probability" for $[0, 1]$-valued continuous events easily follows from the consistency theorem. See Corollary 6.17.

Turning to logic-algebraic independence, the extension of Theorem 4.3 to continuous $[0, 1]$-valued events is given by the following "product law" Theorem 7.18:

> Product is the only continuous $[0, 1]$-valued operation on $[0, 1]^2$ which preserves the consistency of the probabilities (betting odds) of conjunctions of continuous $[0, 1]$-valued events in any two logically independent consistent sets.

Continuing the analysis of probabilistic algebraic consistency, in Theorem 7.13 it is proved that the states of any MV-algebra A can be identified (via an affine isomorphism) with the totality of regular Borel probability measures on the maximal spectral space of A. Then Theorem 7.20 yields the desired symbiosis between finitely additive and countably additive probability:

> Up to isomorphism, the pairs (A, σ) with A an MV-algebra and σ a finitely consistent map (i.e., a state, by Corollary 6.14) of A yield all Kolmogorov triples $(\Omega, \mathcal{B}(\Omega), P)$ for any compact Hausdorff space Ω and regular Borel probability measure P on $\mathcal{B}(\Omega)$. To this purpose, one lets Ω be the maximal spectral space of A, and P be the extension of σ to $\mathcal{B}(\Omega)$ provided by Theorem 7.13.

This result may *mitigate* Kolmogorov's appraisal of his countable additivity axiom as "expedient" and "arbitrary", once we regard his axiom as the natural counterpart of a fundamental consistency notion. This result may also *reconcile* de Finetti finitely additive probability and Kolmogorov countably additive probability, showing that they are the two faces of the same coin.

Let us now consider the following (PSAT_∞) problem: Given a finite set of events with their probabilities/betting odds, *is the consistency of this assignment decidable by a Turing machine*? Using the natural logic-algorithmic machinery of Łukasiewicz logic, Chaps. 8 and 9 will be devoted to proving the decidability of this problem and of its conditional generalization. To specify the input string of symbols in any instance $\mathcal{I}$ of the problem, $\mathcal{I}$ will consist of finitely many events coded by $Ł_\infty$-formulas $\psi_1, \ldots, \psi_m$, and corresponding rational numbers $r_1, \ldots, r_m \in [0, 1]$. We then prove:

> There exists a Turing machine deciding whether a finite list of continuous events ψ_j with rational betting odds r_j is consistent, both in case all events are unconditioned, and in case they are subject to a finite set of conditions, also coded by formulas in Łukasiewicz logic.

These decidability results on probabilistic consistency depend on the interplay between syntactical and semantical consequence in Łukasiewicz logic.

Reverting to boolean algebras and to the multifaceted notion of independence, in Chap. 10 a self-contained proof of de Finetti's exchangeability/representation theorem is given for the countably generated free boolean algebra, and its main consequences. The proof immediately follows from the following result :

> Exchangeable states in the free boolean algebra on countably many generators are limits of product states.

This fundamental result, known as "de Finetti theorem" by antonomasia, is related to the result proved in the first part of the book, concerning the uniqueness of product as a consistency preserving operation for conjunctions of logically independent sets of events.

Part II of the book is written at a level appropriate to upper undergraduate or first year graduate students in mathematics and computer science. As is customary in the study of Łukasiewicz logic and MV-algebras, many results require the synergy of various results in measure theory, polyhedral geometry, lattice-ordered abelian groups, logic, functional analysis, and computation theory. For detailed information on all these classical results used in the book the reader is addressed to the references given at the end of each chapter. On the other hand, for every result on MV-algebras and Łukasiewicz logic needed in the book the reader is addressed, without exception, to the author's joint book with R. Cignoli and I.M.L. D'Ottaviano, "Algebraic Foundations of Many-valued Reasoning", Trends in Logic, volume 7, Kluwer, Dordrecht, (2000), Springer, Dordrecht, (2010), henceforth denoted

“TheBook”,

of which the present book can be considered a continuation focused on the logic-algorithmic aspects of MV-algebraic probabilistic consistency.

I am grateful to the two referees of the original version of this book for their careful and competent reading, detailed corrections, and valuable comments, which resulted in a thoroughly revised version of the original.

I owe a special debt of gratitude to Cecilia for her support and encouragement.

Daniele Mundici

Florence

May 2025

Contents

Part I
Consistency and Independence of {0, 1}-events

In the first two chapters of this book we prepare all necessary material for the proof of the boolean algebraic version de Finetti's consistency theorem, also known as his "Dutch Book Theorem". In Theorem 3.8 it is proved that for any boolean algebra A and finite subset E of A, a map $\beta : E \to [0, 1]$ is consistent, (i.e., β satisfies condition (C) stated in the Introduction), if and only if it is extendable to a state of A. The proof will combine the most basic properties of boolean algebras, (in Chapter 1), with Minkowski's hyperplane separation theorem and "Farkas' lemma", (in Chapter 2). Self-contained proofs of all these results are provided, so that the student may enjoy them on a first reading, or whenever the time seems ripe to her/him.

At the outset of Chapter 3 the probabilistic consistency condition (C) for a map $\beta : E \to [0, 1]$ is elucidated in the framework of a betting game where an ideal bookmaker, named Bookie, is willing to swap his role whenever the smart bettor named Betty so desires. Bookie is "ideal" because the betting odds in his "book" β can be freely used by Betty as the bookmaker, with Bookie in the bettor's role. The inconsistency of Bookie's book means that Betty can devise a bet that will bankrupt Bookie whatever the outcome of the events in E. We will have opportunities to relate the consistency problem to the PSAT problem, a probabilistic generalization of the boolean satisfiability problem SAT of interest in Artificial Intelligence. In Theorem 3 we will prove the boolean algebraic generalization of de Finetti's consistency theorem, also known as his "Dutch Book" theorem. In the bibliographical notes at the end of Chapter 3 we will see that the consistency theorem is nothing else but the Fundamental Theorem of Asset Pricing "without discounting".

Chapter 4 is devoted to the relationships between consistency and the classical probability-free notion of logic-algebraic independence. Theorem 4.3 shows that product is the only consistency preserving operation for the probability of conjunctions of events in two algebraically independent sets. With the exception of the final part of the proof of Theorem 4.3, all proofs in Part I of the book are self-contained.

Combining Theorems 3.2 and 4.3, boolean algebras equipped with a state are shown to naturally correspond to Kolmogorov countably additive probability triples on boolean topological spaces.

The extension of this correspondence to Kolmogorov probability for continuous events on all compact Hausdorff spaces is a main goal of the second part of the book.

All boolean algebras in this book are *proper*, i.e., $0 \neq 1$. All topological spaces have the Hausdorff separation property.

Chapter 1
Finite Boolean Algebras and Their States

On page 307 of his 1931 paper published in volume 17 of *Fundamenta Mathematicae*, de Finetti writes:

> Un evento E è una proposizione, un'affermazione, che non sappiamo ancora se sia vera o falsa
>
> An event E is a proposition, a statement, which we do not yet know whether it is true or false

This expository style is different from the definition-theorem-proof style adopted throughout the present book. The next few pages will be devoted to the *definition* of events and their outcomes in the language of boolean algebras, the *motivation* of these definitions, and the *geometric representation* of events in euclidean finite-dimensional space.

While events in Kolmogorov's axiomatic probability theory are members of boolean algebras closed under countably infinite disjunctions, in de Finetti's probability events are only closed under finite disjunctions. Nevertheless, we will see that these two interpretations of events are essentially equivalent.

Boolean logic is devoted to study simple propositions/events/sets and the compound propositions/events/sets built up from them. Intuitively, a "possible world" ω assigns to each event a yes-no (truth-) value, respecting the syntax of compound events. Thus, e.g., ω assigns to the conjunction of events a and b the minimum of the values assigned by ω to a and b. We come down to earth to give a rigorous treatment of these notions using the terse language of boolean algebras and their homomorphisms.

Boolean algebras provide a rigorous approach to the various operations (typically, of negation, conjunction, disjunction) transforming elementary events into more complicated events. The homomorphisms of boolean algebras into the two-element boolean algebra $\{0, 1\} = \{$"false", "true"$\}$ provide a convenient formal counterpart of the "possible worlds" where the truth-value of these events becomes known. Boolean algebras also provide an algebraic counterpart of "propositions" equipped with the

D. Mundici, *Algebraic Probabilistic Consistency*, Trends in Logic 69,
https://doi.org/10.1007/978-3-031-98337-5_1

connectives of negation, conjunction and disjunction, to speak of complex events in terms of simpler ones. The reader will have various opportunities to see the mutual relationships between these two aspects of boolean algebras.

1.1 Events as Elements of Boolean Algebras

Following standard practice, the mathematical term "iff" stands for "if and only if". Thus for example, an even number is prime iff it equals 2. Sometimes we write $\Leftrightarrow$ instead of "iff". We also use $\Rightarrow$ for "implies". The terms "function", "operation", "map" will be used as synonyms.

Our analysis of events and possible worlds rests upon the following mathematical concepts and their most elementary properties.

A *lattice* is an algebra $L = (L, \wedge, \vee)$ equipped with a partial order such that any two elements $x, y \in L$ have a greatest lower bound (alias the *infimum*, or *meet*) $x \wedge y$ and a least upper bound (also known as the *supremum*, or *join*) $x \vee y$. We say that L is *distributive* if for all $x, y, z \in L$,

$$x \wedge (y \vee z) = (x \wedge y) \vee (x \wedge z) \text{ and } x \vee (y \wedge z) = (x \vee y) \wedge (x \vee z).$$

The *underlying order* of L is defined by the stipulation $x \leq y$ iff $x \wedge y = x$.

A *boolean algebra* $A = (A, 0, 1, \neg, \wedge, \vee)$ is a distributive lattice $(A, \wedge, \vee)$ with a smallest element 0 and a largest element 1, equipped with an operation $\neg$ (called *complementation*) satisfying $x \wedge \neg x = 0$ and $x \vee \neg x = 1$. An element $b \in A$ is said to be an *atom* if it is a nonzero minimal element, in symbols,

$$b \in \mathsf{at}(A).$$

The *two-element boolean algebra* $\{0, 1\} = (\{0, 1\}, \neg, \wedge, \vee)$ is defined by $\neg 0 = 1$, $\neg 1 = 0$, $a \wedge b = \min(a, b)$, $a \vee b = \max(a, b)$ for all $a, b \in \{0, 1\}$. Throughout Part 1 of this book, "events" will be understood as elements of a boolean algebra A acted upon by the operations of A.

Definition 1.1 A *homomorphism* of a boolean algebra $A = (A, 0, 1, \neg, \wedge, \vee)$ into a boolean algebra $A' = (A', 0', 1', \neg', \wedge', \vee')$ is a function $\eta \colon A \to A'$ that preserves the complemented lattice structure:

$$\eta(\neg x) = \neg' x, \quad \eta(x \wedge y) = \eta(x) \wedge' \eta(y), \quad \eta(x \vee y) = \eta(x) \vee' \eta(y).$$

It follows that $\eta(0) = 0'$ and $\eta(1) = 1'$. An *isomorphism* of A onto A' is a one-one function θ of A onto A' such that both θ and its inverse are homomorphisms.

Elements a, b of a boolean algebra A are said to be *incompatible* if $a \wedge b = 0$.

Until further notice, $A = (A, 0, 1, \neg, \vee, \wedge)$ denotes a finite boolean algebra

This restriction will be removed in Chapter 3, §3.4.

Proposition 1.2 *Let* $x, y \in A$.

(i) If $x \neq 0$ *then* x dominates *some atom* $b \in \mathsf{at}(A)$. *In symbols,* $x \geq b$.

(ii) For every $a \in \mathsf{at}(A)$ *exactly one of* $a \leq x$ *or* $a \leq \neg x$ *holds.*

(iii) $x \leq y$ *iff* $x \wedge \neg y = 0$.

(iv) If $x < y$ *then* $y \wedge \neg x \neq 0$.

(v) $\neg x$ *is the largest element of* A *incompatible with* x. *Symmetrically,* $\neg x$ *is the smallest element* $z \in A$ *such that* $x \vee z = 1$.

(vi) A function $\phi \colon A \to B$ *is an isomorphism of boolean algebras* A *and* B *if and only if* ϕ *is onto* B *and*

$$\text{for all } x, y \in A, \quad x \leq y \;\Leftrightarrow\; \phi(x) \leq \phi(y). \tag{1.1}$$

Proof (i) If x is an atom we are done. If x is not an atom, by definition of minimality there is a nonzero $x_1 < x$. If x_1 is an atom we are done. Otherwise there is nonzero $x_2 < x_1$. Since A is finite, any such chain $x > x_1 > x_2 > \ldots$ will end after finitely many steps with a nonzero minimal element.

(ii) If (absurdum hypothesis) $a \leq x$ and $a \leq \neg x$ then, by definition of infimum, $a \leq x \wedge \neg x = 0$ which contradicts $0 \neq a \in \mathsf{at}(A)$. Thus *at most one* of the inequalities $a \leq x$ and $a \leq \neg x$ holds. To show that *at least one* inequality holds, let us assume $a \not\leq x$, i.e., $a \wedge x < a$, with the intent of proving $a \leq \neg x$. Since a is minimal nonzero, then

$$a \wedge x = 0. \tag{1.2}$$

Furthermore,

$$a \wedge \neg x \neq 0. \tag{1.3}$$

For otherwise, combining $a \wedge \neg x = 0$ with condition (1.2) and the distributivity property of A, we obtain $0 = (a \wedge x) \vee (a \wedge \neg x) = a \wedge (x \vee \neg x) = a \wedge 1 = a$, which is impossible. By (i) and (1.3) there is an atom $b \leq a \wedge \neg x$, i.e., $b \leq \neg x$ and $b \leq a$. It follows that $b = a$, whence $a \leq \neg x$, as desired.

(iii) ($\Rightarrow$) From the assumption $x \wedge y = x$ we have $x \wedge \neg y = x \wedge y \wedge \neg y = 0$. ($\Leftarrow$) We first write $y \vee (\neg y \wedge x) = (y \vee \neg y) \wedge (y \vee x) = 1 \wedge (y \vee x) = y \vee x$. From our standing hypothesis $x \wedge \neg y = 0$ we obtain $y \vee x = y \vee (\neg y \wedge x) = y \vee 0 = y$, whence $y \geq x$.

(iv) By (iii), $x \neq y$ iff either $x \wedge \neg y \neq 0$ or $y \wedge \neg x \neq 0$. Therefore, $x < y$ iff $y \neq x$ and $x \leq y$ iff (either $x \wedge \neg y \neq 0$ or $y \wedge \neg x \neq 0$) and $x \wedge \neg y = 0$ iff $y \wedge \neg x \neq 0$ and $x \wedge \neg y = 0$, which implies $y \wedge \neg x \neq 0$.

(v) Distributivity ensures that the supremum s of (the finite set of) all elements of A incompatible with x is also incompatible with x. Therefore, s is the greatest element of A incompatible with x. In particular, since $\neg x$ is incompatible with x, $s \geq \neg x$. By way of contradiction, assume $s > \neg x$. Then from (iv) we can write $0 \neq s \wedge \neg\neg x = s \wedge x = 0$, which is impossible. So $s = \neg x$. The rest is proved similarly.

(vi) If ϕ is an isomorphism then it is onto B, and both ϕ and its inverse preserve the lattice-order structure, whence (1.1) is satisfied. Conversely, assume $\phi\colon A \to B$ is onto B and satisfies (1.1). It follows that ϕ is one-one. Further, both ϕ and ϕ^{-1} preserve the lattice operations. By the characterization of $\neg x$ in (v), ϕ and ϕ^{-1} also preserve the $\neg$ operation. □

As usual, "□" denotes the end-of-proof symbol.

1.2 Representation of Finite Boolean Algebras

Let $\emptyset$ be shorthand for the empty set. For any boolean algebra A let

$$\mathsf{pow}(\mathsf{at}(A)) \quad \text{(read: "the powerset of } \mathsf{at}(A)\text{")} \tag{1.4}$$

be the boolean algebra of all subsets of $\mathsf{at}(A)$, where $0 = \emptyset$, $1 = \mathsf{at}(A)$ and for all subsets X, Y of $\mathsf{at}(A)$,

$$X \vee Y = X \cup Y, \quad X \wedge Y = X \cap Y, \quad \neg X = \mathsf{at}(A) \setminus X = \text{ the complement of } X.$$

Proposition 1.3 *For any $x \in A$ let $\downarrow x$ be the set of atoms dominated by x, $\downarrow x = \{a \in \mathsf{at}(A) \mid a \leq x\}$. We then have*

(i) $\downarrow y = \emptyset$ iff $y = 0$. If $\downarrow x = \{b_1, \ldots, b_l\} \neq \emptyset$ then $x = b_1 \vee \cdots \vee b_l$.

(ii) The function $\downarrow$ is an isomorphism of A onto $\mathsf{pow}(\mathsf{at}(A))$.

Proof (i) By Proposition 1.2(i), the first statement is trivial. For the second statement, by definition of supremum, $x \geq \bigvee_{i=1}^{l} b_i$. By way of contradiction, assume $x > \bigvee_{i=1}^{l} b_i$. By Proposition 1.2(iv), $x \wedge \neg \bigvee_{i=1}^{l} b_i \neq 0$, whence by Proposition 1.2(i) some atom b is dominated by both x and $\neg \bigvee_{i=1}^{l} b_i$. Since $\{b_1, \ldots, b_l\}$ is the list of all atoms dominated by x, we may safely assume $b = b_1$. Then $b_1 \leq \neg \bigvee_{i=1}^{l} b_i = \bigwedge_{i=1}^{l} \neg b_i$, whence $b_1 \leq \neg b_1$. Thus $b_1 \leq b_1 \wedge \neg b_1 = 0$, which is impossible.

(ii) From (i) we have

$$\text{for any } \{b_1, \ldots, b_l\} \subseteq \mathsf{at}(A), \quad \downarrow (b_1 \vee \cdots \vee b_l) = \{b_1, \ldots, b_l\}. \tag{1.5}$$

Thus the function $\downarrow$ is onto pow(at(A)). In the light of Proposition 1.2(vi), there remains to be proved that for all $x, y \in A$, $x \leq y$ iff $\downarrow x \subseteq \downarrow y$. If $x \leq y$ then every atom dominated by x is also dominated by y, and hence, $\downarrow x \subseteq \downarrow y$. Conversely, if $x \not\leq y$ then $x \wedge \neg y \neq 0$ by Proposition 1.2(iii). Proposition 1.2(i) yields an atom a such that $a \leq x \wedge \neg y$. Since $a \leq \neg y$, then by Proposition 1.2(ii), $a \not\leq y$, whence $a \notin \downarrow y$. Since $a \leq x$, $a \in \downarrow x$. In conclusion, $\downarrow x \neq \downarrow y$. □

Part (ii) in Proposition 1.3 is an example of a "representation theorem". Its role is to give the reader a concrete realization of every finite boolean algebra A as the boolean algebra of all subsets of the set of atoms of A, equipped with union, intersection and complement.

The product order in the boolean algebra $\{0, 1\}^n$. As usual, $\mathbb{R}$ denotes the real line, $\mathbb{R}^2$ the cartesian plane, and $\mathbb{R}^n$ the n-dimensional euclidean space. We let $[0, 1]^n$ denote the *unit n-cube* in $\mathbb{R}^n$. Then $\{0, 1\}^n$ is the set of vertices of $[0, 1]^n$. The *standard basis vectors* of $\mathbb{R}^n$ are denoted $e_1, \ldots, e_n$.

Definition 1.4 The *product order* of the set $\{0, 1\}^n$ is defined by the following stipulation:

$x \leq y$ iff whenever a coordinate x_i of x is 1 then so is the coordinate y_i of y.

Geometrically, $x < y$ means that there is a path from x to y consisting of steps along consecutive edges of the unit cube $[0, 1]^n$, where each step moves away from the origin.

A straightforward verification yields:

Proposition 1.5 *The product order makes* $\{0, 1\}^n$ *into a boolean algebra with bottom element* $0 =$ *the origin in* $\mathbb{R}^n$ *and top element* $1 = (1, \ldots, 1)$. *For any* $x, y \in \{0, 1\}^n$ *and* $i = 1, \ldots, n$ *the* i*th coordinate* $(\neg x)_i$ *of* $\neg x$ *equals* $1 - x_i$. *Further,* $(x \wedge y)_i = \min(x_i, y_i)$ *and* $(x \vee y) = \max(x_i, y_i)$. *The atoms of* $\{0, 1\}^n$ *are the* standard basis vectors $e_1, \ldots, e_n$. □

Theorem 1.6 *(*Geometric representation of any finite boolean algebra A.*) With* $\{a_1, \ldots, a_n\} = \mathsf{at}(A)$ *and* $\{e_1, \ldots, e_n\}$ *the standard basis vectors in* $\mathbb{R}^n$, *let the function* $\iota\colon A \to \{0, 1\}^n$ *be defined by stipulating that for all* $x \in A$,

$$\iota(x) = \bigvee_{i=1,\ldots,n} \{e_i \mid a_i \leq x\} \;=\; \sum_{i=1,\ldots,n} \{e_i \mid a_i \leq x\} \in \mathbb{R}^n. \tag{1.6}$$

Then ι *is an isomorphism of* A *onto* $\{0, 1\}^n$, $\iota\colon A \cong \{0, 1\}^n$.

Proof By (1.5) and Proposition 1.3, the linear independence of the standard basis vectors ensures that ι is one-one. To prove that every $v \in \{0, 1\}^n$ is in the range of ι, let us write

$$v = e_{n_1} + \cdots + e_{n_k} \tag{1.7}$$

for a unique subset $\{e_{n_1}, \ldots, e_{n_k}\}$ of $\{e_1, \ldots, e_n\}$. Let $\{a_{n_1}, \ldots, a_{n_k}\}$ be the corresponding set of atoms of A. Let $x_v \in A$ be defined by $\downarrow x_v = \{a_{n_1}, \ldots, a_{n_k}\}$, as per Proposition 1.3(ii). By (1.6), $\iota(x_v) = v$, which shows that the function ι is onto $\{0, 1\}^n$.

In view of Proposition 1.2(vi) there remains to prove that for all $x, y \in A$ $x \leq y \Leftrightarrow \iota(x) \leq \iota(y)$. By Definition 1.4 and (1.6), for all $i = 1, \dots n$ we have

$$e_i \leq \iota(x) \Leftrightarrow a_i \leq x.$$

As a consequence,

$$\begin{aligned} x \leq y &\Leftrightarrow \downarrow x \subseteq \downarrow y, \text{ by Proposition 1.3(ii)} \\ &\Leftrightarrow \text{from } a \in \mathsf{at}(A) \text{ and } a \leq x \text{ it follows that } a \leq y \\ &\Leftrightarrow \text{from } e \in \mathsf{at}(\{0,1\}^n) \text{ and } e \leq \iota(x) \text{ it follows that } e \leq \iota(y) \\ &\Leftrightarrow \iota(x) \leq \iota(y). \end{aligned}$$

The proof is complete. □

1.3 Geometric Representation of the Homomorphisms of *A* Into {0, 1}

For any (always column) vector $v \in \mathbb{R}^n$ we let v^T denote its transpose. Thus, e.g., if v is the vector $\binom{x}{y}$ then v^T is the one-line two-column matrix (x, y) .

For any vector w in $\mathbb{R}^n$ we let $v^T w$ denote the matrix product of v^T and w, i.e., their scalar product in $\mathbb{R}^n$.

For any two (possibly infinite) sets X and Y, by a *bijection* of X onto Y we mean an injective function $f: X \to Y$ with $f(X) = Y$. Equivalently, we say that f is a *one-one correspondence* between X and Y.

We use the notation

$$\mathsf{hom}(A)$$

for the set of homomorphisms of A into the two-element boolean algebra $\{0, 1\}$. It follows that any $\eta \in \mathsf{hom}(A)$ satisfies the identities

$$\eta(\neg x) = 1 - \eta(x), \quad \eta(x \wedge y) = \min(\eta(x), \eta(y)), \quad \eta(x \vee y) = \max(\eta(x), \eta(y)).$$

Intuitively, the homomorphisms of A into $\{0, 1\} = \{no, yes\}$ are the "possible outcomes" of the "events" (i.e., the elements) of A.

The most general example of a homomorphism of A into $\{0, 1\}$ is given by the following result, which in general fails when A is an infinite boolean algebra:

Corollary 1.7 (Geometric representation of $\mathsf{hom}(A)$) *For any A we have:*

(i) Let the function $a \mapsto \eta_a$ send every atom a into the function $\eta_a : A \to \{0, 1\}$ such that for all $x \in A$, $\eta_a(x) = 1$ iff $a \leq x$. Then $a \mapsto \eta_a$ is a bijection of $\mathsf{at}(A)$ *onto* $\mathsf{hom}(A)$. *The inverse bijection maps any $\eta \in \mathsf{hom}(A)$ to the only atom a_η of A such that $\eta(a_\eta) = 1$.*

(ii) Let $\{e_1, \ldots, e_n\} = \mathsf{at}(\{0,1\}^n)$. *For each* $i = 1, \ldots, n$ *let the function* $\theta_{e_i} \colon \{0,1\}^n \to \{0,1\}$ *be defined by stipulating that for every* $v \in \{0,1\}^n$

$$\theta_{e_i}(v) = 1 \;\text{ iff }\; e_i \leq v \;\text{ in the product order of } \{0,1\}^n. \tag{1.8}$$

Then the function $e_i \mapsto \theta_{e_i}$ *(for each* $i = 1, \ldots, n$*), is a bijection of* $\mathsf{at}(\{0,1\}^n)$ *onto* $\mathsf{hom}(\{0,1\}^n)$. *For any atom* e *of* $\{0,1\}^n$, θ_e *is the only homomorphism of* $\mathsf{hom}(\{0,1\}^n)$ *such that* $\theta_e(e) = 1$. *The inverse bijection maps every* $\theta \in \mathsf{hom}(\{0,1\}^n)$ *to the uniquely determined standard basis vector* $e_\theta \in \mathsf{at}(\{0,1\}^n) \subseteq \mathbb{R}^n$ *such that* $\theta(e_\theta) = 1$.

(iii) For every $v \in \{0,1\}^n$ *and* $i = 1, \ldots, n$,

$$\theta_{e_i}(v) = e_i^T v = i\text{th coordinate of } v. \tag{1.9}$$

Proof (i) For all $x, y \in A$, $\eta_a(x \wedge y) = 1$ iff $x \wedge y \geq a$ iff $x \geq a$ and $y \geq a$ iff $\eta_a(x) = 1$ and $\eta_a(y) = 1$. Thus $\eta_a(x \wedge y) = \eta_a(x) \wedge \eta_a(y)$. Similarly, η_a preserves the $\vee$ operation. Preservation of the $\neg$ operation follows from Proposition 1.2(ii). Therefore, $\eta_a \in \mathsf{hom}(A)$. If a and b are distinct atoms of A then $\eta_a(a) = 1$ and $\eta_b(a) = 0$, because $a \not\geq b$. Thus the function $a \mapsto \eta_a$ is one-one. For any $\eta \in \mathsf{hom}(A)$ let $B_\eta \subseteq \mathsf{at}(A)$ be the set of atoms b such that $\eta(b) = 1$. B_η is nonempty, for otherwise (absurdum hypothesis),

$$1 = \eta(1) = \eta(\textstyle\bigvee_{i=1}^n a_i) = \max\{\eta(a) \mid a \in \mathsf{at}(A)\} = 0,$$

which is impossible. We have just proved that the function $a \mapsto \eta_a$ is onto $\mathsf{hom}(A)$. Finally, B_η cannot contain two distinct atoms a, b. For otherwise, (absurdum hypothesis), Proposition 1.2(ii) yields $b \leq \neg a$, whence $0 = 1 - \eta(a) = \eta(\neg a) \geq \eta(b) = 1$, a contradiction. So precisely one atom a_η belongs to B_η. Evidently, $\eta = \eta_{a_\eta}$.

(ii) By Theorem 1.6, this is the special case of (i) for $A = \{0,1\}^n$.

(iii) For arbitrary $v \in \{0,1\}^n$, let $e_{n_1}, \ldots, e_{n_k} \in \mathbb{R}^n$ be the basis vectors $\leq v$. By Proposition 1.3(i) and (1.7) we have $v = e_{n_1} + \cdots + e_{n_k}$. From (1.8) we obtain

$$\begin{aligned}
\theta_{e_i}(v) = 1 \;&\text{iff }\; e_i \leq v, \quad \text{by (9)}\\
&\text{iff }\; e_i \in \{e_{n_1}, \ldots, e_{n_k}\}\\
&\text{iff }\; e_i^T(e_{n_1} + \cdots + e_{n_k}) = 1\\
&\text{iff }\; e_i^T v = 1.
\end{aligned}$$

This completes the proof. □

1.4 Interlude: A Taste of Finitely Generated Free Boolean Algebras

Fix an integer $n = 1, 2, \ldots$ and a set $G = \{X_1, \ldots, X_n\}$.

Does there exist a largest boolean algebra containing G as a generating set?

Let us toss a coin n times, and record the result ("Head"=1, "Tail"=0) of each toss. Suppose for each $i = 1, \ldots, n,\ X_i$ is shorthand for the event "the result of the ith toss is Head". With $\neg, \wedge, \vee$ the usual connectives of boolean logic, let us consider the 2^n equivalence classes (the *minterms*) of the boolean formulas

$$X_1^{\mathsf{b}_1} \wedge \cdots \wedge X_n^{\mathsf{b}_n} \text{ with } X_i^{\mathsf{b}_i} = X_i \text{ if } \mathsf{b}_i = 1, \text{ and } X_i^{\mathsf{b}_i} = \neg X_i \text{ if } \mathsf{b}_i = 0.$$

Each minterm records the possible outcomes of our n coin tosses. The outcome of any coin toss is "independent" of the outcome of any other toss, in the sense that the occurrence or non-occurrence of X_i has no effect on the occurrence or non-occurrence of X_j, $(i \neq j)$.

Let $\mathfrak{B}_n$ be the set of all boolean *formulas* in the *variables* $X_1, \ldots, X_n$, where two formulas are identified iff they are equivalent: thus for instance, $\neg\neg X_1$ is identified in $\mathfrak{B}_n$ with X_1, $X_1 \wedge X_2$ is identified with $X_2 \wedge X_1$, and $X_1 \vee (X_2 \wedge X_3)$ is identified with $(X_1 \vee X_2) \wedge (X_1 \vee X_3)$.[1] The result is the *free boolean algebra on the free generating set* $G = \{X_1, \ldots, X_n\}$. Alternative equivalent definitions of $\mathfrak{B}_n$ are as follows:

$\{X_1, \ldots, X_n\}$ generates $\mathfrak{B}_n$ and $X_1^{\mathsf{b}_1} \wedge \cdots \wedge X_n^{\mathsf{b}_n} \neq 0$ for all $(\mathsf{b}_1, \ldots, \mathsf{b}_n) \in \{0, 1\}^n$. Equivalently, $\{X_1, \ldots, X_n\}$ generates $\mathfrak{B}_n$ and for every boolean algebra A and *function* $f : G \to A$, f uniquely extends to a *homomorphism* of $\mathfrak{B}_n$ into A.

We can now answer the question at the outset of this chapter:

Proposition 1.8 *$\mathfrak{B}_n$ is the largest possible boolean algebra containing G as a generating set.*

Proof For, if B is another boolean algebra generated by G, then by our last definition of $\mathfrak{B}_n$, the identity function $\epsilon \colon X_i \mapsto X_i$ extends to a homomorphism $\tilde{\epsilon}$ of $\mathfrak{B}_n$ into B. Now, $\tilde{\epsilon}$ is onto B, because the X_i generate B. We conclude that the number of elements of B cannot exceed that of $\mathfrak{B}_n$. □

Proposition 1.9 *The minterms of $\mathfrak{B}_n$ are the 2^n atoms of $\mathfrak{B}_n$. Intuitively, they are the "atomic events" of $\mathfrak{B}_n$.*

Each minterm $t \in \mathfrak{B}_n$ uniquely determines a "possible world" of $\mathfrak{B}_n$, i.e., a homomorphism $\mathfrak{B}_n$ of into $\{0, 1\}$, assigning 1 or 0 to any event $e \in \mathfrak{B}_n$, depending

[1] By assigning 0 or 1 to the variables in all possible ways and working in the two-element boolean algebra $\{0, 1\}$ one obtains a familiar mechanical procedure to check if two formulas are logically equivalent.

on whether e dominates the atom t or e is disjoint from t, as per the alternative in Proposition 1.2(ii).

Conversely, any $\eta \in \mathsf{hom}(\mathfrak{B}_n)$ uniquely determines the atom a_η given by the smallest element $a \in \mathfrak{B}_n$ such that $\eta(a) = 1$. The independence of the events X_i results in the largest set of possible worlds. □

Now suppose $n = 3$ and the variables X_1, X_2, X_3 stand for the following events, where "wins" is shorthand for "wins the next FIFA World Cup":

$$X_1 = \text{Brazil wins}, \quad X_2 = \text{ Spain wins}, \quad X_3 = \text{France wins}.$$

From the rules of the FIFA World Cup it follows that not all 2^3 atomic events of $\mathfrak{B}_3$, coded by the minterms available from the variable X_1, X_2, X_3 can occur. For instance, the atomic event $X_1 \wedge X_2 \wedge \neg X_3$ is impossible, and so is the atomic event $X_1 \wedge X_2 \wedge X_3$. The impossible atomic events are precisely those stating that the number of winners is different from 1. Thus, e.g., the event $\neg X_1 \wedge \neg X_2 \wedge \neg X_3$ is possible, since more than the above three teams take part in the Cup.

As a consequence, the boolean algebra A generated by the events X_1, X_2, X_3 subject to the above conditions is strictly smaller than the free three-generated algebra $\mathfrak{B}_3$, the largest possible boolean three-generated algebra. We construct A by deleting from the set of atoms of $\mathfrak{B}_3$ those which code atomic events forbidden by the rules of the FIFA World Cup. A moment's reflection shows that the surviving four atomic events in A are as follows:

$$X_1 \wedge \neg X_2 \wedge \neg X_3, \neg X_1 \wedge X_2 \wedge \neg X_3, \neg X_1 \wedge \neg X_2 \wedge X_3, \neg X_1 \wedge \neg X_2 \wedge \neg X_3.$$

As expected, the boolean algebra A has $2^4 = 16$ elements/events, fewer than the $2^8 = 256$ elements of $\mathfrak{B}_3$. Beyond the events X_1, X_2, X_3, and the four atoms of A, examples of events of A include $\neg X_1 \wedge X_2$ (which in A is the same as X_2), the impossible event $X_1 \wedge X_2$ (i.e., the zero element of A), the sure event $X_1 \vee \neg X_1$ (i.e., the top element 1 of A), and a few others.

As in Proposition 1.3(i), each event of A is the disjunction of the atomic events it dominates. Thus, e.g., in A we have

$$\neg X_1 = (\neg X_1 \wedge \neg X_2 \wedge X_3) \vee (\neg X_1 \wedge X_2 \wedge \neg X_3) \vee (\neg X_1 \wedge \neg X_2 \wedge \neg X_3),$$

making $\neg X_1$ equal to the supremum of three atoms of A. By contrast, in the free algebra $\mathfrak{B}_3$ the element $\neg X_1$ is the supremum of four atoms.

By Proposition 1.3(ii), up to isomorphism, every n-generated boolean algebra arises from a similar procedure reducing the atoms of a free boolean algebra $\mathfrak{B}_n$.

These reductions procedures and their generalizations will be reconsidered in Chapters 3, 9, and 10.

1.5 States of Boolean Algebras

Definition 1.10 A *state* of a boolean algebra A is a function $\sigma\colon A \to [0,1]$ with $\sigma(1) = 1$, having the *additivity property*: For all $x, y \in A$, if $x \wedge y = 0$ then $\sigma(x \vee y) = \sigma(x) + \sigma(y)$.

An easy verification shows that *every homomorphism of A into $\{0,1\}$ is a state.*

As the reader will recall, a (possibly infinite) set X in euclidean space $\mathbb{R}^n$ is *convex* if for any two points $x, y \in X$ the segment joining x and y is contained in X. For any set $\{x_1, \ldots, x_l\} \subseteq \mathbb{R}^n$, the *convex hull* $\mathsf{conv}(x_1, \ldots, x_l)$ is the set of all *convex combinations* of $x_1, \ldots, x_l$, i.e., all points x of $\mathbb{R}^n$ of the form

$$x = \lambda_1 x_1 + \cdots + \lambda_n x_n, \quad 0 \le \lambda_i \in \mathbb{R}, \ (i = 1, \ldots, n), \quad \sum_{i=1}^{n} \lambda_i = 1.$$

An element $z \in X$ is *extreme* (in X) if whenever $z \in \mathsf{conv}(a, b)$ for some $a, b \in X$ then $z = a$ or $z = b$.

The geometric representation of every state σ of a boolean algebra A is the object of the following corollary.

Corollary 1.11 (Representation of the states of a finite boolean algebra)
Let σ be a state of the boolean algebra $\{0,1\}^n$. Let the closed convex set $\mathsf{S}(\{0,1\}^n) \subseteq \mathbb{R}^n$ and the vector $v_\sigma \in \mathbb{R}^n$ be respectively defined by

$$\mathsf{S}(\{0,1\}^n) = \mathsf{conv}(e_1, \ldots, e_n) \quad \textit{and} \quad v_\sigma = (\sigma(e_1), \ldots, \sigma(e_n)). \tag{1.10}$$

Then

(i) $(v_\sigma)_1 + \cdots + (v_\sigma)_n = 1$. *Thus* $v_\sigma \in \mathsf{S}(\{0,1\}^n)$.
(ii) *For all* $v = (v_1, \ldots, v_n) \in \{0,1\}^n$, $\sigma(v) = v_\sigma^T v$.
(iii) v_σ *is the only vector in* $\mathbb{R}^n$ *such that for all* $v \in \{0,1\}^n$, $\sigma(v) = v_\sigma^T v$. *Therefore, any state is uniquely determined by the values it assigns to the atoms* e_i.
(iv) *The map* $\tau \mapsto v_\tau$ *is a bijection of the set of states of the boolean algebra* $\{0,1\}^n$ *onto the set* $\mathsf{conv}(e_1, \ldots, e_n) \subseteq \mathbb{R}^n$.

Proof (i) The top element 1 in the boolean algebra $\{0,1\}^n$ is the sum $e_1 + \cdots + e_n = e_1 \vee \cdots \vee e_n$. Since $e_i \wedge e_j = 0$ for $i \neq j$ then 1 is the supremum of incompatible elements of $\{0,1\}^n$. By additivity, $1 = \sigma(\bigvee_{i=1}^n e_i) = \sum_{i=1}^n \sigma(e_i)$, whence $v_\sigma \in \mathsf{S}(\{0,1\}^n)$.

(ii) For each $i = 1, \ldots, n$ let us agree to write $0e_i =$ the origin in $\mathbb{R}^n$ and $1e_i = e_i$. The vector v is a linear combination of the pairwise orthogonal vectors $e_1, \ldots, e_n$ with uniquely determined coefficients 1 or 0 according as $e_i \le v$ or $e_i \not\le v$. Further, v is the supremum of the e_i with the same coefficients. By Corollary 1.7(iii), $v_i = \theta_{e_i}(v) \in \{0,1\}$, whence

$$v = \sum_{i=1}^{n} \theta_{e_i}(v)e_i = \bigvee_{i=1}^{n} \theta_{e_i}(v)e_i.$$

By (1.10) and the additivity property of σ we can write

$$\begin{aligned}\sigma(v) &= \sigma\left(\bigvee_{i=1}^{n} \theta_{e_i}(v)e_i\right) = \sum_{i=1}^{n} \sigma(\theta_{e_i}(v)e_i) = \sum_{i=1}^{n} \theta_{e_i}(v)\sigma(e_i) \\ &= \sum_{i=1}^{n} v_i(v_\sigma)_i = v^T v_\sigma = v_\sigma^T v.\end{aligned}$$

(iii) Suppose $w \in \mathbb{R}^n$ satisfies $w^T v = \sigma(v)$ for all $v \in \{0, 1\}^n$. Then by (i)–(ii), for all $i = 1, \ldots n$, we have $(w^T - v_\sigma^T)e_i = 0$ whence $w_i = (v_\sigma)_i$ and $w = v_\sigma$.

(iv) From (i)–(iii). □

1.6 De Finetti's "Fundamental Theorem of Probability"

On page 246 of Chapter XVI of his "Investigation of the laws of thought", Boole writes:

> "the object of the theory of probabilities might be thus defined. Given the probabilities of any events, of whatever kind, to find the probability of some other event connected with them."

In the paper where he proved his consistency theorem de Finetti gave an operational criterion for the probability of the new event to exist.

For a self-contained proof of this theorem we prepare the following notation and terminology, which will also be used in a later chapter:

A *hyperplane* H in euclidean space $\mathbb{R}^n$ is a subset of $\mathbb{R}^n$ of the form $\{x \in \mathbb{R}^n \mid l^T x = w\}$ for some nonzero vector $l \in \mathbb{R}^n$ and real number w. The sets $\{x \in \mathbb{R}^n \mid l^T x \geq w\}$ and $\{x \in \mathbb{R}^n \mid l^T x \leq w\}$ are the two *(closed) halfspaces* with boundary H.

Theorem 1.12 (De Finetti's "Fundamental Theorem of Probability") *Let A be a boolean algebra with n atoms and state space $\mathsf{S}(A)$. Fix a finite subset $E = \{h_1, \ldots, h_m\}$ of A along with a function $\beta \colon E \to [0, 1]$, and let*

$$S_\beta = \{\sigma \in \mathsf{S}(A) \mid \sigma \supseteq \beta\} = \textit{be the set of states of } A \textit{ extending } \beta.$$

Next, for any $h \in A$ let

$$S_\beta(h) = \{\sigma(h) \mid \sigma \in S_\beta\} \tag{1.11}$$

be the set of possible values assigned to h by all states extending β. *Then* $S_\beta(h)$ *is either empty, or is a closed interval contained in* $[0, 1]$ *possibly consisting of a single point.*

Proof In view of Theorem 1.6, we can identify A with the boolean algebra $\{0, 1\}^n$. Then E is a set of vertices $h_j \in \{0, 1\}^n$ of the n-cube $[0, 1]^n$. For all $\sigma \in \mathsf{S}(A)$ and $w \in \{0, 1\}^n$, the value $\sigma(w)$ is given by the scalar product $v_\sigma^T w$, with $v_\sigma = (\sigma(e_1), \ldots, \sigma(e_n))$ the vector associated to σ by the geometric representation in Corollary 1.11. In particular, v_σ lies in the closed convex set

$$\mathsf{S}(\{0, 1\}^n) = \mathsf{conv}(e_1, \ldots, e_n), \text{ with } e_1, \ldots, e_n \text{ the standard basis vectors of } \mathbb{R}^n.$$

For every $j = 1, \ldots, m$, the set $\{\sigma \in \mathsf{S}(\{0, 1\}^n) \mid \sigma(h_j) = \beta(h_j)\}$ of states of $A = \{0, 1\}^n$ extending the singleton function $h_j \mapsto \beta(h_j)$ is the set of vectors

$$\{v_\sigma \in \mathsf{S}(\{0, 1\}^n) \mid \sigma(h_j) = \beta(h_j)\} = \{v_\sigma \in \mathsf{S}(\{0, 1\}^n) \mid v_\sigma^T h_j = \beta(h_j)\}.$$

This is the intersection of $\mathsf{S}(\{0, 1\}^n)$ with the hyperplane $H_j \subseteq \mathbb{R}^n$ of all vectors $w \in \mathbb{R}^n$ such that $w^T h_j = \beta(h_j)$. As a consequence, the set $S_\beta \subseteq \mathsf{S}(\{0, 1\}^n)$ of states of A extending β coincides with $\mathsf{S}(\{0, 1\}^n) \cap H_1 \cap \cdots \cap H_m$. It follows that S_β is a closed convex (a fortiori connected) subset of $\mathsf{S}(\{0, 1\}^n)$. The set $S_\beta(h)$ in (1.11) has the form $\{v_\sigma^T h \in [0, 1] \mid \beta \subseteq \sigma \in \mathsf{S}(\{0, 1\}^n)\}$. Thus $S_\beta(h)$ is the range of the continuous $[0, 1]$-valued function $v_\sigma \mapsto v_\sigma^T h$, where the vector v_σ ranges over the closed connected space S_β. Elementary topology shows that $S_\beta(h)$ is closed and connected. In conclusion, if nonempty, $S_\beta(h)$ is a closed (possibly a singleton) interval in $[0, 1]$. □

Historical and bibliographical notes. De Finetti's quotation at the outset of this chapter can be found in [3, §7].

For Boole the atoms of the free boolean algebra $\mathfrak{B}_n$ are its "constituents". De Finetti says that each atomic event is a "case". Kolmogorov calls each atom an "elementary event". As will be shown in Theorem 4.6, the 2^n atoms of $\mathfrak{B}_n$ correspond to the points (also known as "sample points") of the Stone dual space of $\mathfrak{B}_n$. An infinite boolean algebra A need not have atoms, but the set $\mathsf{hom}(A)$ conveniently replaces the set of atoms of A in any case. For more information on free boolean algebras see [12, §9] or [16, §14]. In [1, Chapter XVIII] Boole gives examples of probabilistic inference problems, together with a detailed discussion of their solutions. Also see [14, 2.8]. In [2, p. 91] Boole describes inconsistent situations, where the solution does not exist. The proof that the set of possible probabilities of the new event in Boole's problem, if nonempty, is a closed interval contained in [0, 1] was given by de Finetti in [4, p. 13]. Also see [5, 3.10.1, p. 112], and [6]. In Finetti's paper [4, p. 9], Boole is cited in connection with the problem of giving a criterion for the probability of the new event ϕ_{m+1} to be *uniquely determined* by the known probabilities of the events $\phi_1, \ldots, \phi_m$. It is noted that the maximum number of "possible cases" (=Boole's "constituents") is at most 2^N (with N the number of variables), because some constituents may turn out to be "impossible".

As in Boole's problem, let us suppose that elements $h_1, \ldots, h_m$ of a boolean algebra A are assigned probabilities $p_1 = \beta(h_1), \ldots, p_m = \beta(h_m) \in [0, 1]$, and h is a new element. De Finetti's solution of Boole's problem yields a closed interval $I \subseteq [0, 1]$ such that the set of possible probabilities of h coincides with I. I is nonempty if and only if β is consistent. Now fix $j = 1, \ldots, m$ and suppose the probability p_j of h_j undergoes a small perturbation, $p_j \mapsto p_j + \mathrm{d}p_j$. The paper [14] analyzes the resulting modification of the vertices of I. For instance, it is proved that the partial derivative $\frac{\partial\,\mathrm{length}(I)}{\partial\beta(h_j)}(p_j)$ has a rational value for any $0 < p_j < 1$. If each probability $\beta(h_1), \ldots \beta(h_m)$ is rational and each event is coded by a boolean formula, then all these partial derivatives are rational and effectively computable.

Boole's remarks on the object of the theory of probabilities are taken up today in the PSAT (probabilistic satisfiability) problem and its optimization version. Here events $h_1, \ldots, h_m$ are assigned probabilities $p_1, \ldots, p_m$, and one has to specify the possible values of the probability p_{m+1} of a new event h_{m+1}. For computational purposes, all events p_i are coded by special boolean *formulas* $\psi_1, \ldots, \psi_m$ (known as clauses), and all $p_1, \ldots, p_m$, are *rational* numbers. When there is no possible value for p_{m+1} the problem is said to be *inconsistent*, using the same terminology adopted throughout this book. If the problem is consistent, Theorem 1.12 shows that the set of possible values of p_{m+1} is a closed (possibly a singleton) interval contained in $[0, 1]$. To provide a fundamental necessary and sufficient condition for this set to be nonempty, it will take de Finetti's consistency (also known as "Dutch Book") Theorem 3.2.

PSAT is equivalent to the problem of checking the consistency of the input $\{(\psi_1, p_1), \ldots, (\psi_m, p_m)\}$. It is a generalization of the satisfiability problem SAT for boolean formulas, and has the same computational complexity as SAT. PSAT and its optimization version pertain to a vibrant research area in various domains, including defeasible reasoning, automated deduction, formal epistemology, and uncertainty management. For more information, also including de Finetti's contributions to Boole's problem and its modern reformulation, see [8, 11].

States of boolean algebras are also known as "measures". Theorem 4.10 below yields an affine isomorphism of the space of states of every boolean algebra A onto the space of regular Borel probability measures on the Stone space of A. The terminology "state" adopted throughout this book avoids confusion between the finite additivity of states of the algebra A and the countable additivity of Borel probability measures on the Stone space of A.

Our choice of the specific term "state" rests on the categorical equivalence Γ between MV-algebras and lattice-ordered abelian groups with a distinguished order unit, for short, *unital ℓ-groups*. Γ will be introduced in Chapter 6, §6.2. By definition, the states of any unital ℓ-group (G, u) are unit-preserving monotone homomorphisms into $(\mathbb{R}, 1)$. As we will see in Theorem 6.10(iii), the state space of (G, u) is affinely homeomorphic to the state space of the MV-algebra $A = \Gamma(G, u)$. Boolean algebras are just idempotent MV-algebras.

The states of quantum systems described by C*-algebras are deeply related to the states of the partially ordered abelian groups associated to them. As a notable example, Elliott classification [7] and Grothendieck K_0 functor [9] yield a bijection of the set of tracial states of any Bratteli approximately finite-dimensional C*-algebra $\mathfrak{A}$ (AF-algebra, for short) onto the set of states of the partially ordered unital group $(K_0(\mathfrak{A}), K_0(\mathfrak{A})^+, [1_{\mathfrak{A}}])$. See [10, p. 310 and references therein]. When the Murray-von Neumann order of $\mathfrak{A}$ is a lattice, $\mathfrak{A}$ is said to be an AFℓ-algebra. Then the map $\mathfrak{A} \mapsto \Gamma(K_0(\mathfrak{A}))$ yields a bijection of all isomorphism classes of AFℓ-algebras onto isomorphism classes of countable MV-algebras. The tracial states of $\mathfrak{A}$ are in one-one correspondence with the states of $\Gamma(K_0(\mathfrak{A}))$. See [13, 15] for details.

References

1. Boole, G. (1958). *An investigation of the laws of thought, on which are founded the mathematical theories of logic and probabilities, Walton and Maberley, London, (1854).* NY: Reprinted by Dover.
2. Boole, G. (1854). XIII. On the conditions by which solutions of questions in the theory of probabilities are limited. The London, Edinburgh and Dublin Philosophical Magazine and Journal of Science, 8(50), 91–98. doi: 10.1080/14786445408651898.
3. de Finetti, B. (1931). Sul significato soggettivo della probabilità. Fundamenta Mathematicae, 17, 298–329. Translated into English as: On the Subjective Meaning of Probability. In P. Monari & D. Cocchi (Eds.), Probabilitá e Induzione, Clueb, Bologna (pp. 291–321) (1993). Reprinted in [6, pp. 191–222].
4. de Finetti, B. (1937). La prévision: ses lois logiques, ses sources subjectives. Annales de l'Institut H. Poincaré, 7, 1–68 English translation by Henry E. Kyburg Jr., as: Foresight: Its Logical Laws, its Subjective Sources. In: H. E. Kyburg Jr., H. E. Smokler, Studies in Subjective Probability, J. Wiley, New York, pp. 93–158, (1964). Second edition published by Krieger, New York, pp. 53–118, (1980). Reprinted in [6, pp. 335–400].
5. de Finetti, B. (1974). Theory of probability (Vol. 1). Wiley.
6. de Finetti, B. (2006). Opere Scelte. (Selected Works) (Vol. 1). Unione Matematica Italiana, Edizioni Cremonese, Firenze.
7. Elliott, G. A. (1976). On the classification of inductive limits of sequences of semisimple finite-dimensional algebras. *Journal of Algebra, 38*(1), 29–44.
8. Georgakopoulos, G., Kavvadias, D., & Papadimitriou, C. H. (1988). Probabilistic satisfiability. *Journal of Complexity, 4*, 1–11.
9. Goodearl, K. R. (1982). Notes on real and complex C*-algebras, Shiva mathematics series (Vol. 5). Shiva Publishing Ltd.
10. Goodearl, K. R. (1986). Partially ordered abelian groups with interpolation. Mathematical Surveys and Monographs, Vol 20. American Mathematical Society.
11. Hansen, P., & Jaumard, B. (2000). Probabilistic satisfiability. In J. Kohlas & S. Moral (Eds.), Handbook of defeasible reasoning and uncertainty management systems (Vol. 5, pp. 321–367). Springer.
12. Koppelberg, S. (1989). General theory of Boolean algebras. In Handbook of Boolean algebras (Vol. 1). Edited by J.D. Monk with the cooperation of R. Bonnet, Elsevier, Amsterdam.
13. Mundici, D. (1986). Interpretation of AF C^*-algebras in Łukasiewicz sentential calculus. *Journal of Functional Analysis, 65*, 15–63.
14. Mundici, D. (2021). The differential of probabilistic entailment. Annals of Pure and Applied Logic, 172, Article 102945.
15. Mundici, D. (2023). AF-algebras with lattice-ordered K_0: Logic and computation. Annals of Pure and Applied Logic, 174(1), Article 103182.
16. Sikorski, R. (1960), Boolean algebras. In Ergebnisse der Mathematik und ihrer Grenzgebiete (3rd ed., Vol. 25), Springer.

Chapter 2
Minkowski Hyperplane Separation, Farkas' Lemma

One easily proves that two disjoint closed disks in $\mathbb{R}^2$ are separated by a line. With relatively little extra effort one can prove that two disjoint closed convex sets in $\mathbb{R}^3$ are separated by a plane, and generalize this to any two disjoint convex closed sets in $\mathbb{R}^n$. However, this is not what we need for the proof of Farkas' lemma, a key tool for the proof of the probabilistic consistency theorem in the next chapter. Here, one of the two disjoint convex sets to be separated is not closed. What we need is Lemma 2.5, a particular case of Minkowski's hyperplane separation theorem. A small number of preliminary elementary facts about convex sets in euclidean n-space will be proved in the next few pages, culminating in a self-contained proof of Farkas' lemma.

2.1 Minkowski Hyperplane Separation

Lemma 2.1 *The closure* $\mathsf{cl}\, X$ *of a convex set* $X \subseteq \mathbb{R}^n$ *is convex.*

Proof We may assume $X \neq \emptyset$. For any $x, y \in \mathsf{cl}\, X$ we must prove that the segment $[x, y] = \{(1-\lambda)x + \lambda y \mid \lambda \in [0, 1]\}$ is contained in $\mathsf{cl}\, X$. By definition of $\mathsf{cl}\, X$, the set $X \subseteq \mathbb{R}^n$ contains converging sequences $x_n \to x$ and $y_n \to y$. By the assumed convexity of X, for all n we have the inclusion $[x_n, y_n] \subseteq X$. The sequence of midpoints m_n of $[x_n, y_n]$ converges to the midpoint m of $[x, y]$. Being the limit of a convergent sequence of points of X, m lies in $\mathsf{cl}\, X$. More generally, for each $\lambda \in [0, 1]$ the sequence $(1-\lambda)x_n + \lambda y_n$ converges to $(1-\lambda)x + \lambda y$. We have just proved that $(1-\lambda)x + \lambda y$ lies in $\mathsf{cl}\, X$ for every $\lambda \in [0, 1]$. In other words, $[x, y] \subseteq \mathsf{cl}\, X$, as desired to complete the proof. □

Let us recall that the scalar product of two (always column) vectors $u, v \in \mathbb{R}^n$ is given by matrix product $u^T v$, with u^T the transpose of u.

For $w \in \mathbb{R}^n$ we let $|w|$ denote its euclidean norm, (or "length") $(w^T w)^{1/2}$.

D. Mundici, *Algebraic Probabilistic Consistency*, Trends in Logic 69,
https://doi.org/10.1007/978-3-031-98337-5_2

Lemma 2.2 *Let $X \subseteq \mathbb{R}^n$ be a nonempty closed convex set.*

(i) X has a unique shortest *vector, i.e., a vector x_* such that*

$$|x_*| = \inf\{|x| \mid x \in X\}. \tag{2.1}$$

(ii) Let $t \in \mathbb{R}^n$. Then there is a unique point u in X such that $|u - t| = \inf\{|x - t| \mid x \in X\}$.

(iii) Assume $0 \notin X$. Let x_* be the shortest vector in X given by (i). Then for all $y \in X$, $x_*^T y > |x_*|^2/2$.

(iv) Let x_* be the shortest vector in X. Assume $0 \notin X$. Then the hyperplane

$$H = \{z \in \mathbb{R}^n \mid x_*^T z = |x_*|^2/2\}$$

strongly separates *the origin from X. (In other words, the distance of the origin from H is > 0, the distance from H of every $y \in X$ is > 0, and X and the origin are contained in opposite closed halfspaces with boundary H.)*

(v) More generally, for any nonempty closed convex set $Y \subseteq \mathbb{R}^n$ and point $e \in \mathbb{R}^n \setminus Y$, letting x_* be the unique point of Y closest to e, we have $(x_* - e)^T(x - e) > |x_* - e|^2/2$ for each $x \in Y$. In other words, the hyperplane

$$K = \{z \in \mathbb{R}^n \mid (x_* - e)^T(z - e) = |x_* - e|^2/2\}$$

strongly separates e from Y.

Proof (i) If the origin 0 belongs to X we have nothing to prove. Otherwise, let ξ be shorthand for $\inf\{|x| \mid x \in X\}$. It is no loss of generality to assume that X is bounded. By definition of the infimum in (2.1) there is a sequence of points $z_m \in X$ with $|z_m| - \xi < 1/m$ and with a convergent subsequence z_{m_i}. The point $z_* = \lim z_{m_i}$ belongs to X, because X is closed. Since $z \mapsto |z|$ is a continuous real-valued function, then $\xi = \lim |z_{m_i}| = |\lim z_{m_i}| = |z_*|$. To prove the uniqueness of z_* suppose (absurdum hypothesis) $z^* \in X$ is different from z_* and has the same length as z_*. The triangle with vertices $0, z_*, z^*$ is isosceles. The midpoint of the interval $[z_*, z^*]$ has a distance $< |z_*|$ from 0, which is impossible.

(ii) The translated set $X - t = \{x - t \mid x \in X\}$ is also nonempty closed and convex. Distances $|a - b|$ are preserved under translation. Now apply (i).

(iii) Arguing by way of contradiction, let us assume that there exists $y \in X$ such that

$$x_*^T y \leq |x_*|^2/2. \tag{2.2}$$

The closed interval $[x_*, y] = \{(1 - \kappa)x_* + \kappa y \mid \kappa \in [0, 1]\}$ is contained in X, and for all $\kappa \in [0, 1]$ we have:

$$|x_*|^2 \leq |(1 - \kappa)x_* + \kappa y|^2, \quad \text{since } x_* \text{ is the shortest vector in } X$$

$$\begin{aligned} &= (1-\kappa)^2|x_*|^2 + \kappa^2|y|^2 + 2(1-\kappa)\cdot\kappa\, x_*^T y \\ &\le |x_*|^2 + \kappa^2|x_*|^2 - 2\kappa|x_*|^2 + \kappa^2|y|^2 + (1-\kappa)\kappa|x_*|^2, \text{ by (2.2)}. \end{aligned}$$

Therefore,

$$\begin{aligned} 0 &\le \kappa^2|x_*|^2 - 2\kappa|x_*|^2 + \kappa^2|y|^2 + \kappa|x_*|^2 - \kappa^2|x_*|^2 \\ &= -\kappa|x_*|^2 + \kappa^2|y|^2. \end{aligned}$$

We then have $|x_*|^2 \le \kappa|y|^2$ for each $\kappa > 0$. It follows that $|x_*|^2 = 0$, whence x_* coincides with the origin of $\mathbb{R}^n$, a contradiction.

(iv) Immediate from (iii). □

Lemma 2.3 (Supporting hyperplane theorem) *Let X be a convex subset of $\mathbb{R}^n$ and b a boundary point of X. Then there is a hyperplane H containing b such that X is contained in one of the two closed halfspaces bounded by H. H is known as a* supporting hyperplane *for X at b.*

Proof By definition of boundary, b lies in the closure $\mathsf{cl}\, X$ of X in $\mathbb{R}^n$, and there is a sequence of points $b_n \notin \mathsf{cl}\, X$ with $b_n \to b$. By Lemma 2.1 $\mathsf{cl}\, X$ is convex. For each $n = 1, 2, \ldots$, let $c_n \in \mathsf{cl}\, X$ be the closest point to b_n as given by Lemma 2.2(ii). Let υ_n be the unit vector $(c_n - b_n)/|c_n - b_n|$. By Lemma 2.2(iv)–(v),

$$\frac{|c_n - b_n|}{2} < \upsilon_n^T(x - b_n) \quad \text{for each } x \in X, \text{ in particular for } x = b. \tag{2.3}$$

The bounded sequence υ_n has a convergent subsequence, say, converging to the unit vector υ. Then by (2.3),

$$\lim \frac{|c_n - b_n|}{2} \le \lim \upsilon_n^T(b - b_n) = \upsilon^T \lim(b - b_n) = 0,$$

whence

$$\lim \frac{|c_n - b_n|}{2} = 0. \tag{2.4}$$

From (2.3)–(2.4), for each $x \in X$ we have

$$0 \le \lim \upsilon_n^T(x - b_n) = \upsilon^T(x - b), \quad \text{whence } \upsilon^T b \le \upsilon^T x.$$

Thus the hyperplane $H = \{x \in \mathbb{R}^n \mid \upsilon^T x = \upsilon^T b\}$ supports X at b. □

Corollary 2.4 *Let $X \ne \mathbb{R}^n$ be a nonempty closed convex subset of $\mathbb{R}^n$. Then X is the intersection of all halfspaces containing it. Actually, X is the intersection of all* supporting halfspaces *of X, i.e. those halfspaces which are the boundary of some supporting hyperplane of X.* □

We are now ready to prove the specific hyperplane separation lemma needed for the proof of de Finetti's consistency theorem.

Lemma 2.5 (Particular case of Minkowski's hyperplane separation theorem) *Let* $O \subseteq \mathbb{R}^n$ *be the set of all vectors whose coordinates are* < 0. *We say that* O *is the* south-west open octant. *Let* R *be a linear subspace of* $\mathbb{R}^n$ *disjoint from* O. *Then for some vector* $u \in \mathbb{R}^n$ *the hyperplane* $H = \{x \in \mathbb{R}^n \mid u^T x = 0\}$ *has the following (separation) property for* O *and* R:

$$R \subseteq H, \quad (\textit{i.e.}, u^T r = 0 \textit{ for all } r \in R).$$

and

$$u^T y < 0 \textit{ for all } y \in O, \textit{ whence } H \cap O \textit{ is empty}.$$

Proof The set $R + O = \{r + o \in \mathbb{R}^n \mid r \in R, \ o \in O\}$ is convex and does not contain the origin. Its closure $\mathsf{cl}(R + O)$ is convex, by Lemma 2.1. Since R and O are disjoint, the origin is a boundary point of $\mathsf{cl}(R + O)$. Next, Lemma 2.3 provides a supporting hyperplane H for $\mathsf{cl}(R + O)$ at 0. Since R is linear and $0 \in R \cap H$ then R is contained in H. By construction, O is contained in one of the two *open* halfspaces with boundary H. The other open halfspace contains a vector u orthogonal to H such that $u^T y < 0$ for all $y \in O$. Since $R \subseteq H$, for any such u we automatically have $u^T r = 0$ for all $r \in R$. □

2.2 Farkas' Lemma

Our next goal is to prove Farkas' lemma, the last basic ingredient of de Finetti's consistency theorem, which will find use also in the second part of the book. This is a relevant example of a "theorem of the alternative", or an "*aut-aut* theorem", where for two conditions C_1 and C_2 is it shown:

—it is never the case that both C_1 and C_2 hold;

—it is always the case that either C_1 or C_2 holds.

In other words, this fact can be expressed by saying that C_1 is logically equivalent to the negation of C_2.

Lemma 2.6 (Farkas, Gordan, Stiemke, et al.) *Let* M *be a real matrix with* n *rows and* m *columns. Then precisely one of the following conditions holds:*

(i) There is a (column) vector $s \in \mathbb{R}^m$ *such that every coordinate of the vector* $Ms \in \mathbb{R}^n$ *is* < 0.

(ii) There is a nonzero vector $u \in \mathbb{R}^n$ *such that every coordinate of* u *is* ≥ 0 *and* $u^T M = 0 =$ *the origin in* $\mathbb{R}^m$. *(As above,* u^T *is the transpose of* u.*)*

Proof Conditions (i) and (ii) are promptly seen to be incompatible. For, if both are assumed to hold we have the contradiction

$$0 \neq \underbrace{u^T}_{u\neq 0,\text{ each coordinate}\geq 0} \underbrace{(Ms)}_{\text{each coordinate}<0} = (u^T M)s = 0. \tag{2.5}$$

There remains to prove that if (i) fails then (ii) holds. Failure of (i) means that the range R of the linear operator $M\colon \mathbb{R}^m \to \mathbb{R}^n$ is disjoint from the south-west open octant $O = \{v \in \mathbb{R}^n \mid \text{ each coordinate of } v \text{ is } < 0\}$. R is a linear subspace of $\mathbb{R}^n$ containing the origin. The hyperplane separation Lemma 2.5 now yields a nonzero column vector $u \in \mathbb{R}^n$ and a hyperplane H such that

$$R \subseteq H = \{x \in \mathbb{R}^n \mid u^T x = 0\} \quad \text{and} \quad O \subseteq \{x \in \mathbb{R}^n \mid u^T x < 0\}.$$

The latter inclusion implies that all coordinates of u are ≥ 0. Evidently, $u^T M = 0$. For otherwise, (absurdum hypothesis), there is a vector $q \in \mathbb{R}^m$ with $(u^T M)q \neq 0$. As a member of R, the vector Mq is orthogonal to u^T, whence $u^T(Mq) = 0$, a contradiction. We have thus proved that if (i) fails then (ii) holds. □

Historical and bibliographical notes. Minkowski's hyperplane separation theorem goes back to [3, §§2–3]. See [1, 2, 4, 6] for (the multifarious variants of) what is generally known as Farkas' lemma. For their applications see, e.g., [5, §7.3] and [7, §1.4].

References

1. Farkas, G. (1902). Theorie der Einfachen Ungleichungen. *Journal für die reine und angewandte Mathematik, 124*, 1–27. https://doi.org/10.1515/crll.1902.124.1
2. Gordan, P. (1873). Über die Auflösung linearer Gleichungen mit reellen cofficienten. Mathematische Annalen, 6, 23–28. https://link.springer.com/article/10.1007/BF01442864.
3. Minkowski, H. (1911). Theorie der konvexen Körper, insbesondere Begründung ihres Oberflächenbegriffs. In Gesammelte Abhandlungen von Hermann Minkowski (Bd 2). Teubner, Leipzig. Reprinted Chelsea Publishing Co., New York, pp. 131–229 (1967).
4. Perng, C. (2017). On a class of theorems equivalent to Farkas's lemma. Applied Mathematical Sciences, 11(44), 2175–2184. https://doi.org/10.12988/ams.2017.77218.
5. Schrijver, A. (1986). *Theory of linear and integer programming*. Wiley.
6. Stiemke, E. (1915). Über positive Lösungen homogener linearer Gleichungen. *Mathematische Annalen, 76*, 340–342. https://doi.org/10.1007/BF01458147
7. Ziegler, G. M. (1995). Lectures on polytopes. "Graduate Text in Mathematics", Vol. 152 Springer.

Chapter 3
De Finetti's Consistency ("Dutch Book") Theorem

On page 308 of his 1931 paper in Fundamenta Mathematicae, de Finetti writes (in his own italics):

> Un individuo è coerente nel valutare le probabilità di certi eventi se qualunque gruppo di puntate $S_1, S_2, \ldots, S_n$ un competitore faccia su un insieme qualunque di eventi $E_1, E_2, \ldots, E_n$ fra quelli che egli ha considerato, non è possibile che il guadagno G del competitore risulti *in ogni caso positivo*.
>
> (An individual is consistent in assessing the probabilities of certain events if for any set of bets $S_1, S_2, \ldots, S_n$ a competitor places on any set of events $E_1, E_2, \ldots, E_n$ among those he has considered, it is not possible for the competitor's G gain to be *positive in any case*.)

In the language of boolean algebras (for the "events") and their homomorphisms (for their "possible worlds"), de Finetti's notion of probabilistic "consistency" will be reproduced below in Definition 3.1. To better understand the nature of this fundamental notion we will preliminarily embed the above terse presentation into a game theoretic context.

3.1 A Game Theoretic Interpretation of Consistency

Bookie (the "individual") is an *ideal* bookmaker. Today he sells bets on events $h_1, \ldots, h_m$ dealing with the possible outcomes of forthcoming football matches. Each h_j is assigned a "betting odd" $\beta(h_j) \in [0, 1]$. The finite set of pairs

$$\{(h_1, \beta(h_1)), \ldots, (h_m, \beta(h_m))\}$$

is Bookie's "book". Betty (the "competitor") is a clever bettor. If $\beta(h_j)$ is fair for her, she now pays Bookie $s_j \cdot \beta(h_j)$ (euro, for definiteness), with the understanding that

D. Mundici, *Algebraic Probabilistic Consistency*, Trends in Logic 69,
https://doi.org/10.1007/978-3-031-98337-5_3

Bookie will pay her back the full "bet" (also known as the "stake") s_j if h_j occurs.[1] For definiteness, throughout we will make the following stipulation:

$$\text{All financial transactions are given the Betty-to-Bookie orientation.} \tag{3.1}$$

If Betty thinks $\beta(h_j)$ is excessive, why not bet a *negative* s_j? What better proof that $\beta(h_j)$ is Bookie's "ideal" betting odd for h_j? Bookie's acclaimed status as an ideal bookmaker makes him available to swap roles with Betty: He then accepts to pay Betty $|s_j| \cdot \beta(h_j)$. Betty will pay him $|s_j|$ if h_j occurs.

As de Finetti argues on page 309 of his paper cited above, each betting odd $\beta(h_j)$ (the "probability" of h_j in de Finetti's terms) must lie in the unit interval $[0, 1]$. For otherwise, if $\beta(h_j) > 1$, letting Z denote one zillion, Betty's bet s_j of $-Z$ will result in Betty's net profit $Z \cdot (\beta(h_j) - 1) > 0$ if h_i occurs, and $Z \cdot (\beta(h_j) - 0)$ otherwise. Needless to say, if $\beta(h_j) < 0$, things are going to get even worse for Bookie: If Betty bets Z, then he has to pay her $Z \cdot |\beta(h_j)|$, with the prospect of paying an additional Z euro if h_i occurs. Therefore, first of all, consistency requires that $0 \leq \beta(h_j) \leq 1$ for each $j = 1, \ldots, m$.

Which other conditions must the betting odds $\beta(h_1), \ldots, \beta(h_m)$ satisfy to avoid "inconsistency", i.e., Bookie's bankruptcy, "whatever the outcome of the events" $h_1, \ldots, h_m$?

To take care mathematically of the innocent looking expressions "whatever the outcome of the events" and "in any possible world", at the outset of Chap. 1 we resorted to boolean algebras and their homomorphisms into the two-element boolean algebra. Definition 3.1 and de Finetti's Theorem 3.2 will help us to delve deeper into algebraic probabilistic consistency.

3.2 Probabilistic Consistency for Finite Boolean Algebras

We now come to the fundamental notion of a consistent assignment of real numbers in $[0, 1]$ to "events" $h_1, \ldots h_m$ understood as elements of a finite boolean algebra A. The finiteness hypothesis for A will be dropped in Sect. 3.4.

Recall from Corollary 1.7(i) the bijection of the set $\{a_1, \ldots, a_n\} = \mathsf{at}(A)$ onto the set $\{\eta_1, \ldots, \eta_n\} = \mathsf{hom}(A)$ of homomorphisms of A into the two-element boolean algebra $\{0, 1\}$.

Definition 3.1 Let $E = \{h_1, \ldots h_m\}$ be a subset of a boolean algebra A and $\beta\colon E \to [0, 1]$ a function. Then β is said to be *inconsistent* if

$$\text{for some } s_1, \ldots, s_m \in \mathbb{R}, \ \sum_{j=1}^{m} s_j(\beta(h_j) - \eta_i(h_j)) < 0 \ \text{ for all } \ i = 1, \ldots, n. \tag{3.2}$$

[1] Needless to say, Bookie will pay back 0 if h_j does not occur.

If β is not inconsistent we say it is *consistent*.

Understanding Definition 3.1. Let us agree to say that β is a "book". The real numbers $s_1, \ldots, s_m$ will be indifferently called (Betty's) "bets", or "stakes". For any $s_j \in \mathbb{R}$, the balance of the single bet s_j on event h_j is given by $s_j(\beta(h_j) - \eta_i(h_j))$ in the "possible world" $\eta_i \in \mathsf{hom}(A)$.[2] Since zero bets are possible, we may assume Betty is betting on all events $h_1, \ldots, h_m$ in the book. The total balance of her bet in η_i is then

$$\sum_{j=1}^{m} s_j(\beta(h_j) - \eta_i(h_j)). \tag{3.3}$$

By definition, Bookie's book β is inconsistent if Betty can devise positive or negative bets $s_1, \ldots, s_m \in \mathbb{R}$ that guarantee her a minimum profit of 1, equivalently, one zillion, "regardless of the outcome of the events h_j", i.e., in any "possible world" $\eta_i \in \mathsf{hom}(A)$. Correspondingly, Bookie has a sure loss of at least one (zillion euro).

The totality of "possible outcomes", or "possible worlds", is made precise by the homomorphisms $\eta_1, \ldots, \eta_n \in \mathsf{hom}(A)$. For a panoramic overview of all events h_j, $(j = 1, \ldots, m)$ and all possible worlds η_i, $(i = 1, \ldots, n)$, we prepare the following $n \times m$ matrix M:

$$M = \begin{pmatrix} \beta(h_1) - \eta_1(h_1) & \beta(h_2) - \eta_1(h_2) & \ldots & \beta(h_m) - \eta_1(h_m) \\ \beta(h_1) - \eta_2(h_1) & \beta(h_2) - \eta_2(h_2) & \ldots & \beta(h_m) - \eta_2(h_m) \\ \ldots & \ldots & \ldots & \ldots \\ \beta(h_1) - \eta_n(h_1) & \beta(h_2) - \eta_n(h_2) & \ldots & \beta(h_m) - \eta_n(h_m) \end{pmatrix} \tag{3.4}$$

Let $s_1, \ldots, s_m \in \mathbb{R}$ be Betty's bets on $h_1, \ldots, h_m$ respectively. Fix a row $i = 1, \ldots, n$ and a column $j = 1, \ldots, m$. In case $\eta_i(h_j) = 0$, i.e., if h_j does not occur in the possible world η_i, then the term $M_{i,j}$ in M coincides with $\beta(h_j)$. Recalling our orientation convention (3.1), Bookie's profit/loss is $s_j \cdot \beta(h_j) \in \mathbb{R}$. On the other hand, in case h_j occurs in the possible world η_i then $M_{i,j} = \beta(h_j) - 1$ and Bookie's profit/loss is $s_j \cdot (\beta(h_j) - 1)$. In either case, Bookie's profit/loss if Betty bets s_j on the event h_j is equal to $s_j \cdot (\beta(h_j) - \eta_i(h_j))$ in the possible world η_i.

Let $s \in \mathbb{R}^m$ be the (column) vector whose coordinates are $s_1, \ldots, s_m$. Then the vector $Ms \in \mathbb{R}^n$ is shorthand for the total balance (3.3) in all possible worlds $\eta_1, \ldots, \eta_n$. Bookie's β is inconsistent iff Betty can devise a bet $s \in \mathbb{R}^n$ such that every coordinate of Ms is < 0. Equivalently, $\sum_{j=1}^{m} s_j(\beta(h_j) - \eta_i(h_j)) < 0$ for every $i = 1, \ldots, n$.

[2] It is understood that the value $\eta_i(h_j)$ is 1 if h_j occurs, and is 0 if h_j does not occur in (the possible world) η_i, $i = 1, \ldots, n$.

3.3 De Finetti's Consistency Theorem for Finite Boolean Algebras

Theorem 3.2 (De Finetti's consistency theorem for finite A) *For any boolean algebra A with atoms $a_1, \ldots, a_n$ and corresponding homomorphisms $\eta_1, \ldots, \eta_n \in \mathsf{hom}(A)$, let $E = \{h_1, \ldots, h_m\} \subseteq A$ and $\beta\colon E \to [0, 1]$. Then precisely one of the following conditions holds:*

(i) β is inconsistent in the sense of Definition 3.1. In other words, there is a column vector $s = (s_1, \ldots, s_m) \in \mathbb{R}^m$ such that every coordinate of the vector $Ms \in \mathbb{R}^n$ is < 0. In symbols,

$$\sum_{j=1}^{m} s_j(\beta(h_j) - \eta_i(h_j)) < 0 \ \textit{for every} \ \ i = 1, \ldots, n.$$

(ii) β is extendable to a state σ of A.

Stated otherwise, β is consistent in A if and only if it is extendable to a state of A.

Proof In view of Theorem 1.6, upon identifying A with the boolean algebra $\{0, 1\}^n$ of the vertices of the n-cube $[0, 1]^n$, each atom a_i of A is realized as the ith standard basis vector e_i in $\mathbb{R}^n$. Furthermore, for each $j = 1, \ldots, m$ the element h_j is realized as a vertex of the n-cube $[0, 1]^n$. Since the isomorphism ι of Theorem 1.6 now coincides with identity, by Corollary 1.7(ii) we can write

$$\eta_i = \theta_i = \theta_{e_i} \in \mathsf{hom}(\{0, 1\}^n) \text{ for each } i = 1, \ldots, n. \tag{3.5}$$

Proof that if condition (i) fails then condition (ii) holds. Suppose condition (i) fails, i.e., β is consistent. By Farkas' Lemma 2.6 there is a nonzero column vector $(u_1, \ldots, u_n) \in \mathbb{R}^n$ with each coordinate $u_i \geq 0$ and $u^T M = 0 =$ the origin in $\mathbb{R}^m$. Without loss of generality, $u_1 + \cdots + u_n = 1$, and we may write

$$u^T M = 0, \quad 0 \leq u_i \ \ (i = 1, \ldots, n), \quad u_1 + \cdots + u_n = 1. \tag{3.6}$$

For each $j = 1, \ldots, m$ the jth column C_j of the $n \times m$ matrix M in (3.4) has the form $C_j = C_{j,\text{left}} - C_{j,\text{right}}$, where

$$C_{j,\text{left}} = \underbrace{(\beta(h_j), \ldots, \beta(h_j))}_{n \text{ terms}}. \tag{3.7}$$

As for the column vector $C_{j,\text{right}}$, for each $i = 1, \ldots, n$ its ith coordinate $(C_{j,\text{right}})_i$ satisfies the identities

$$(C_{j,\text{right}})_i = \eta_i(h_j) = \theta_i(h_j) = e_i^T h_j = i\text{th coordinate of } h_j \in \{0, 1\}^n. \tag{3.8}$$

This follows from 1.6-1.7 and (3.5). By (3.6), $u^T M = 0$, i.e., $u^T C_{j,\text{right}} = u^T C_{j,\text{left}}$ for each $j = 1, \ldots, m$. By (3.7)-(3.8), for each $j = 1, \ldots, m$ we can write

$$\beta(h_j) = u^T C_{j,\text{right}} = u^T h_j. \tag{3.9}$$

Let the state σ of the boolean algebra $\{0, 1\}^n$ be defined by

$$\sigma(e_i) = u_i, \quad (i = 1, \ldots, n).$$

By Corollary 1.11(iii) and (3.9), $\sigma(h_j) = u^T h_j = \beta(h_j)$ for each $j = 1, \ldots, m$. This shows that σ extends β, whence condition (ii) holds.

Proof that conditions (i) and (ii) are incompatible. By way of contradiction, assume that both conditions (i) and (ii) hold. Let the state σ of A extend β, as per condition (ii). Recalling Corollary 1.11, let the vector $v_\sigma \in \mathsf{S}(\{0, 1\}^n)$ be defined by $v_\sigma^T w = \sigma(w)$ for all $w \in \{0, 1\}^n$. In particular, $v_\sigma^T h_j = \sigma(h_j)$ for each $j = 1, \ldots, m$. Since all coordinates of v_σ are ≥ 0 and their sum is equal to 1, then

$$v_\sigma^T C_{j,\text{left}} = \beta(h_j) \quad \text{for all} \quad j = 1, \ldots, m.$$

By (3.8) we also have $v_\sigma^T C_{j,\text{right}} = v_\sigma^T h_j = \sigma(h_j)$. Since $\sigma \supseteq \beta$ then

$$v_\sigma^T C_{j,\text{left}} = v_\sigma^T C_{j,\text{right}} \quad \text{for all} \quad j = 1, \ldots, m, \quad \text{whence } v_\sigma^T M = 0.$$

Since also condition (i) is assumed to hold, there is a vector $s = (s_1, \ldots, s_m) \in \mathbb{R}^m$ such that every coordinate of the vector $Ms \in \mathbb{R}^n$ is < 0.

As in the proof of Farkas' Lemma, conditions (i) and (ii) now jointly imply the contradiction

$$0 \neq \underbrace{v_\sigma^T}_{v_\sigma \in \mathsf{S}(\{0,1\}^n)} \quad \underbrace{(M(s_1, \ldots, s_m))}_{s_j < 0, \text{ for each } j=1,\ldots,m} = (v_\sigma^T M)(s_1, \ldots, s_m) = 0.$$

We have just shown that conditions (i) and (ii) are incompatible.

This completes the proof. □

Remark 3.3 Readers who surf the internet and are interested in football betting odds will often come across tables like Table 3.1 below:

For instance the number 4.50 in the first row means that if team A wins you will win 4.50 euro for each euro you bet on this result. This is just the reciprocal of Bookie's betting odd, stating that you have to pay 1/4.50 = for each euro you'll receive if A wins. Traditionally, "1" represents a home win, "X" stands for a draw, and "2" represents an away team win. One can easily verify that all books specified by the numbers on the five lines of this table are inconsistent—with one exception, due to the present author. One may then wonder/guess why the table only displays *reciprocals* of betting odds, and why inconsistency has no negative effect on non-ideal bookmakers.

Table 3.1 Betting odds for next Sunday's football matches. A selection

MATCH	1	X	2
A vs. B	4.50	3.90	1.70
C vs. D	2.49	3.09	2.78
E vs. F	1.10	6.35	12
G vs. H	1.25	10	10
I vs. J	2.50	2.50	4

3.4 De Finetti's Consistency Theorem for All Boolean Algebras

Generalizing Theorem 3.2, in Theorem 3.8 we will show that our restriction to finite boolean algebras is inessential. To make sure that the reader does not miss it, we enclose our general stipulation in a box:

A henceforth stands for an arbitrary boolean algebra.

Definition 3.4 As above, $\mathsf{hom}(A)$ denotes the set of homomorphisms of A into the two-element boolean algebra $\{0, 1\}$. Also, a *state* σ of A is a $[0, 1]$-valued function on A which is additive on disjunctions of incompatible elements of A and satisfies $\sigma(1) = 1$.

Lemma 3.5 *Let B be a* finite *boolean algebra and B' a subalgebra of B.*
(i) The homomorphisms of B' into $\{0, 1\}$ are precisely the restrictions to B' of the homomorphisms of B into $\{0, 1\}$.
(ii) The states of B' are precisely the restrictions to B' of the states of B.

Proof (i) Let $\epsilon \in \mathsf{hom}(B)$. Trivially, the restriction $\epsilon \restriction B'$ of ϵ to B' is a homomorphism of B' into $\{0, 1\}$.

Conversely, let $\eta \in \mathsf{hom}(B')$, with the intent of extending η to some $\eta^* \in \mathsf{hom}(B)$. By Corollary 1.7(i) we may write $\eta = \eta_e$ for a unique atom e of B'. By Proposition 1.3, each atom a of B' is a nonzero element of B, and dominates a nonempty set $[a]$ of atoms of B. If $a \neq b \in \mathsf{at}(B')$ then $[a] \cap [b] = \emptyset$. (For otherwise, some atom c of B is dominated by both a and b, whence c is dominated by $a \wedge b = 0$, which is impossible.) As a consequence, upon writing

$$c \approx d \ \text{ iff, } c \text{ and } d \text{ are dominated by the same atom } a \text{ of } B', \ (c, d \in \mathsf{at}(B)),$$

we obtain an equivalence relation $\approx$ over $\mathsf{at}(B)$. The function $a \mapsto [a],\ a \in \mathsf{at}(B')$ maps $\mathsf{at}(B')$ onto the set of $\approx$-equivalence classes. Furthermore, for each $x \in B'$ letting $a_1, \ldots, a_l$ be the atoms of B' dominated by x, we have the identity

$$x = \bigvee\{b \in \mathsf{at}(B) \mid b \in ([a_1] \vee \cdots \vee [a_l])\}. \tag{3.10}$$

Turning to our homomorphism $\eta_e \in \mathsf{hom}(B')$, let us arbitrarily pick an atom $e^* \in [e]$, and let $\eta_{e^*} \in \mathsf{hom}(B)$ be the corresponding homomorphism. By a final application of Corollary 1.7, for every $y \in B$, $\eta_{e^*}(y) = 1$ if and only if y dominates e^*. An easy verification using (3.10) shows that the homomorphism $\eta^* = \eta_{e^*}$ is an extension of η_e.

(ii) The restriction to B' of any state of B is a state of B'. Conversely, let τ be a state of B'. With the notation of (i), for each $a \in \mathsf{at}(B')$ arbitrarily pick an atom $a^* \in [a] \subseteq \mathsf{at}(B)$. Let the function $\tau^\circ\colon \mathsf{at}(B) \to [0, 1]$ be defined by

$$\tau^\circ(a^*) = \tau(a) \text{ for each } a \in \mathsf{at}(B'), \text{ and } \tau^\circ(b) = 0 \text{ for all other atoms of } B.$$

By Corollary 1.11(iii), any state of the finite boolean algebra B is uniquely determined by the values it gives to the atoms of B. It is now easy to verify that τ° uniquely determines a state τ^* of B which extends τ. □

Lemma 3.6 *Let B' be a* finite *subalgebra of an* infinite *boolean algebra A.*

(i) The states of B' are precisely the restrictions to B' of the states of A.

(ii) The homomorphisms of B' into $\{0, 1\}$ are precisely the restrictions to B' of the homomorphisms of A into $\{0, 1\}$.

(iii) Every boolean algebra A has a homomorphism into $\{0, 1\}$ and a state.

Proof (i) Let σ be a state of B'. By Lemma 3.5(ii), for every *finite* subalgebra B of A containing B', σ is extendable to a state of B. Let the set $\tilde{S}_{\sigma,B} \subseteq [0, 1]^A$ be defined by $\tilde{S}_{\sigma,B} = \{f : A \to [0, 1] \mid f \restriction B \text{ is a state of } B \text{ extending } \sigma\}$. By definition of the product topology of $[0, 1]^A$, $\tilde{S}_{\sigma,B}$ is a nonempty closed subset of the Tychonov cube $[0, 1]^A$. For any finite family of finite subalgebras $B_1, \ldots, B_u$ of A containing B' the intersection I of the closed sets $\tilde{S}_{\sigma,B_1}, \ldots, \tilde{S}_{\sigma,B_u}$ is nonempty. As a matter of fact, letting B_{u+1} be the subalgebra of A generated by the finite set $B_1 \cup \cdots \cup B_u$, it follows that B_{u+1} is finite and the set I contains the nonempty set $\tilde{S}_{\sigma,B_{u+1}}$. Next let $S^* = \bigcap\{\tilde{S}_{\sigma,B} \mid B \text{ a finite subalgebra of } A \text{ containing } B'\}$. From the compactness of the Tychonov cube $[0, 1]^A$ it follows that S^* is nonempty. Any element of S^* is a state of A extending σ.

(ii) Follows from a routine variant of the proof of (i) using Lemma 3.5(i).

(iii) This is immediate from (i)-(ii), since $\mathsf{hom}(B') \neq \emptyset$. □

In general, the set $\mathsf{hom}(A)$ is no longer indexed by the atoms of the boolean algebra A. Suffice to say that an infinite boolean algebra A need not have atoms. Thus the definition of consistency gets the following form, which is equivalent to Definition 3.1 for finite boolean algebras, and encompasses the general case when the boolean algebra A is infinite.

Definition 3.7 Let $E = \{h_1, \ldots h_m\}$ be a subset of a boolean algebra A and $\beta\colon E \to [0, 1]$ a function. Then β is said to be *inconsistent* (in A) if

$$\text{there are } s_1, \ldots, s_m \in \mathbb{R} \text{ such that } \sum_{j=1}^{m} s_j(\beta(h_j) - \eta(h_j)) < 0 \text{ for any } \eta \in \mathsf{hom}(A). \tag{3.11}$$

If β is not inconsistent we say it is *consistent.*

Theorem 3.2 has the following generalization to every (finite or infinite) boolean algebra A:

Theorem 3.8 (De Finetti consistency theorem for any boolean algebra)
Let E be a finite subset of a boolean algebra A and $\beta\colon E \to [0, 1]$ a function. Then β is consistent if and only if it is extendable to a state of A.

Proof Let $E = \{h_1, \ldots, h_m\}$. The subalgebra B_E of A generated by E is finite. (Indeed, as noted in Sect. 1.4 of Chap. 1, the number of elements of B_E is $\leq 2^m$.)

($\Rightarrow$) For any $s\colon E \to \mathbb{R}$ let us agree to say that $\eta_s \in \mathsf{hom}(B_E)$ is a *witness of the s-consistency of β in B_E* if

$$\sum_{j=1}^{m} s(h_j)(\beta(h_j) - \eta_s(h_j)) \geq 0. \tag{3.12}$$

Since, by assumption, β is consistent in the ambient algebra A, for every $t\colon E \to \mathbb{R}$ there is a witness $\theta_t \in \mathsf{hom}(A)$ of the t-consistency of β in A,

$$\sum_{j=1}^{m} t(h_j)(\beta(h_j) - \theta_t(h_j)) \geq 0.$$

A fortiori, the restriction of θ_t to B_E is a witness of the t-consistency of β in B_E. Since t is any arbitrary real-valued function defined on E, β is consistent in B_E. By Theorem 3.2, β is extendable to a state τ of B_E. By Lemma 3.6(i), τ is extendable to a state ρ of A, whence $\beta \subseteq \tau$ is extendable to ρ.

($\Leftarrow$) Let σ be a state of A extending β. The restriction $\sigma \upharpoonright B_E$ is a state of B_E extending β. By Theorem 3.2, β is consistent in B_E. There remains to prove that β is consistent in A. Arguing by way of contradiction, let us assume the existence of a function $s\colon E \to [0, 1]$ such that

$$\sum_{j=1}^{m} s(h_j)(\beta(h_j) - \eta(h_j)) < 0 \quad \text{for every } \eta \in \mathsf{hom}(A).$$

Intuitively, s is Betty's ruinous bet for Bookie. By Lemma 3.6(ii),

$$\sum_{j=1}^{m} s(h_j)(\beta(h_j) - \theta(h_j)) < 0 \quad \text{for every } \theta \in \mathsf{hom}(B_E).$$

By Definition 3.7, β is inconsistent in B_E, a contradiction. □

Corollary 3.9 *For any boolean algebra A, the finite restrictions of states of A are the same as the consistent maps in A.* □

Corollary 3.10 *A function $f: A \to [0, 1]$ is a state of the boolean algebra A if and only if f is* finitely consistent, *in the sense that every finite restriction of f is consistent in A.*

Proof For any state σ of A, every finite restriction ζ of σ is extendable to the state σ. By Theorem 3.8, ζ is consistent. We conclude that σ is finitely consistent.

Conversely, let $f: A \to [0, 1]$ be a finitely consistent map. By definition, every finite restriction r of f is consistent. By Theorem 3.8, r is extendable to a state σ_r of A. By direct inspection, the set S_r of states σ extending r is a closed subset of the space $\mathsf{S}(A) \subseteq [0, 1]^A$ of states of A. Let $\{r_1, \ldots, r_u\}$ be a finite but otherwise arbitrary set of finite restrictions of f. Then it is easy to see that the closed $S_{r_1} \cap \cdots \cap S_{r_u}$ is nonempty. This is usually referred to as the "Finite Intersection Property" of the family $\{S_r \mid r \text{ a finite restriction of } f\}$. Since the set $\mathsf{S}(A)$ of states is compact the intersection $S^* = \bigcap\{S_r \mid r \text{ a finite restriction of } f\}$ is nonempty. Actually, S^* is a singleton, whose only element is f and is a state of A. □

The proof of the above corollary yields the following more general result:

Corollary 3.11 *A* $[0, 1]$*-valued map f defined on a (possibly infinite) subset of a boolean algebra A is extendable to a state of A iff so is every finite restriction of f iff every finite restriction of f is (probabilistically) consistent.* □

Corollary 3.12 (Gödel compactness theorem, general form) *A* $\{0, 1\}$*-valued map g defined on a (possibly infinite) subset of a boolean algebra A is extendable to a homomorphism of A into* $\{0, 1\}$ *iff so is every finite restriction r of g iff r is logically consistent.* □

Remark 3.13 From Lemmas 3.5–3.6 and Theorem 3.8 it follows that a function $\beta: E \to [0, 1]$ is consistent in a boolean algebra $A \supseteq E$ iff it is consistent in the algebra B_E generated by E in A iff it is consistent in any algebra A^* containing A as a subalgebra. Hence the consistency of β is largely indifferent to the chosen boolean algebra $A \supseteq E$. However, the specification of *some* ambient boolean algebra A for the set E of "events" and their operations is necessary to give a meaning to such expressions as "in any possible case", or "in any possible world", which occur in the definition of consistency and in the consistency theorem.

3.5 Logical Consistency as a Special Case of Probabilistic Consistency

This section requires some familiarity with boolean propositional logic. The first chapters of any book on mathematical logic will provide all necessary background. For details see the final section on bibliographical notes.

As in of Chap. 1 §1.4, let $\mathfrak{B}_n$ be the free boolean algebra over the free generating set $\{X_1, \ldots, X_n\}$. Elements of $\mathfrak{B}_n$ are boolean formulas ϕ in the variables $X_1, \ldots, X_n$ up to logical equivalence. The boolean operations naturally act on $\mathfrak{B}_n$ as the boolean connectives act on formulas. Thus, e.g., let $\phi(X_1, \ldots, X_n)$ be a formula and $[\phi] \in \mathfrak{B}_n$ be the (infinite!) set of formulas logically equivalent to ϕ. Then for any formula ψ logically equivalent to ϕ the equivalence class $[\neg\psi]$ of $\neg\psi$ coincides with the complement $\neg[\phi] \in \mathfrak{B}_n$ of the equivalence class $[\phi]$. Any formula ψ logically equivalent to ϕ is said to *code* the element $[\phi]$ of $\mathfrak{B}_n$.

For each $i = 1, \ldots, n$, X_i is a special kind of a formula, known as a *variable*. Traditionally, the same notation is used for X_i and for the equivalence class $[X_i] \in \mathfrak{B}_n$. Likewise, $\neg, \wedge, \vee$ denote both the connectives acting on formulas, and the corresponding operations of the boolean algebra $\mathfrak{B}_n$.

A *truth-valuation* is a $\{0, 1\}$-valued function v defined on all formulas $\psi = \psi(X_1, \ldots, X_n)$, having the following properties:

$$v(\neg\psi) = 1 - v(\psi), \quad v(\psi \wedge \phi) = \min(v(\psi), v(\phi)), \quad v(\psi \vee \phi) = \max(v(\psi), v(\phi)).$$

In other words, v is a $(\neg, \wedge, \vee)$-preserving map of the (*absolutely free*) algebra of formulas into the two-element boolean algebra $\{0, 1\}$. The non-ambiguity of the syntax ensures that v is uniquely determined by the values it assigns to the variables $X_1, \ldots, X_n$. By setting $\eta_v([\phi]) = v(\phi)$, v uniquely determines a homomorphism $\eta_v \in \mathsf{hom}(\mathfrak{B}_n)$. Conversely, every $\eta \in \mathsf{hom}(\mathfrak{B}_n)$ has the form $\eta = \eta_w$ for a unique truth-valuation w.

As we have seen in the first lines of the Introduction, when the function β of Definition 3.1 takes values in the two-element set $\{0, 1\}$ rather than in $[0, 1]$, de Finetti consistency coincides with consistency in boolean logic. In more detail:

Corollary 3.14 *Let $E = \{h_1, \ldots, h_m\}$ be a finite subset of the free boolean algebra $\mathfrak{B}_n$. Let β be a* $\{0, 1\}$-valued *function defined on E. Then the following conditions are equivalent:*

(i) For each $j = 1, \ldots, m$ let $h_j \in E$ be coded by a formula ϕ_j in the variables $X_1, \ldots, X_n$. With b_j shorthand for $\beta(h_j)$, let $\phi_j^{\mathsf{b}_j} = \phi_j$ if $\mathsf{b}_j = 1$ and $\phi_j^{\mathsf{b}_j} = \neg\phi_j$ if $\mathsf{b}_j = 0$, it follows that $\phi_1^{\mathsf{b}_1} \wedge \cdots \wedge \phi_m^{\mathsf{b}_m}$ is satisfiable, *in the sense that some truth-valuation assigns the constant value 1 to each formula $\phi_j^{\mathsf{b}_j}$.*

(ii) β is extendable to a homomorphism η of $\mathfrak{B}_n$ into $\{0, 1\}$.

(iii) β is extendable to a state *of $\mathfrak{B}_n$.*

(iv) β is consistent *in the sense of Definition 3.7, i.e.,*

$$\text{for any } s_1, \ldots, s_m \in \mathbb{R} \text{ there is } \eta \in \mathsf{hom}(\mathfrak{B}_n) \text{ such that } \sum_{j=1}^{m} s_j(\beta(h_j) - \eta(h_j)) \geq 0.$$

Proof (i)⇔(ii) For each $j = 1, \ldots, m$, let $h_j^{\mathsf{b}_j} = h_j$ if $\mathsf{b}_j = 1$ and $h_j^{\mathsf{b}_j} = \neg h_j$ if $\mathsf{b}_j = 0$. By hypothesis, some truth-valuation v assigns the constant value 1 to each formula

$\phi_j^{\mathsf{b}_j}$. It follows that $\eta(h_1^{\mathsf{b}_1}) = \cdots = \eta(h_m^{\mathsf{b}_m}) = 1$ for $\eta = \eta_v \in \mathsf{hom}(\mathfrak{B}_n)$. Therefore, $\eta(h_j) = \beta(h_j)$ for each $j = 1, \ldots, m$, and η extends β. To prove the converse, use the fact that every $\eta \in \mathsf{hom}(\mathfrak{B}_n)$ has the form $\eta = \eta_w$ for a unique truth-valuation w.

(ii)⇔(iv) Assuming (ii), let $\eta^* \in \mathsf{hom}(\mathfrak{B}_n)$ be an extension of β. Then the consistency condition (iv) trivially holds for η constantly equal to η^*.

Conversely, assuming that (ii) does not hold, Betty can devise the following strategy to bankrupt Bookie and show that (iv) does not hold:

— She bets $s_j = 7$ (zillion euro) on each event $h_j \in E$ such that $\beta(h_j) = 0$. Accordingly, she now pays $0 = 7 \cdot \beta(h_j)$ to Bookie, in the hope of winning 7 in case (i.e., in the "possible world" $\eta \in \mathsf{hom}(\mathfrak{B}_n)$ where) h_j occurs. Note that this partial bet does not involve any financial risk for Betty.

— To complete her bet, she bets $s_i = -7$ on each event h_i such that $\beta(h_i) = 1$. Recalling our stipulation about the Betty-to-Bookie orientation of all financial transactions in this game, it is Bookie who now pays Betty $7 = 7 \cdot \beta(h_i) = |s_i \beta(h_i)|$. She will pay him back $7 = |s_i \eta(h_i)|$ in the possible world η where h_i occurs. Also this partial bet does not involve any financial risk for Betty.

Our assumption that (ii) does not hold means that in any possible world $\eta \in \mathsf{hom}(\mathfrak{B}_n)$ some event $h_\eta \in E$ satisfies the inequality $\eta(h_\eta) \neq \beta(h_\eta)$. Therefore, Betty's total bet $s_1, \ldots, s_m$ is not only risk-free, but she will also make a net profit of at least 7 (zillion) in any possible world η. Recalling (3.1) we can write

$$\sum_{j=1}^{m} s_j(\beta(h_j) - \eta(h_j)) \leq -7 < 0,$$

which shows that (iv) also fails with Bookie.

(iii)⇔(iv) This is the content of Theorem 3.8. □

Using the fact that every boolean algebra A is the homomorphic image of some free boolean algebra, with little extra effort Corollary 3.14 can be shown to hold for any boolean algebra A.

Remark 3.15 One might argue that every book must assign probabilities to formulas rather than to elements of a boolean algebra. Thus for instance, inputs for the PSAT problems are special boolean formulas, known as "clauses". This will also be our attitude later in this book when dealing with decidability problems. However, for our present purposes it is enough to note that if Bookie assigns different probabilities $p_1 < p_2$ to two logically equivalent formulas ϕ_1, ϕ_2, then Betty can bankrupt him by fixing a very large bet s and then buying ϕ_1 for $s \cdot p_1$ and selling ϕ_2 for $s \cdot p_2$. In any case Betty's enormous net profit will be $s \cdot (p_2 - p_1)$.

This shows that, insofar as we are concerned with the consistency problem, we may assume that the set of events in Bookie's book is a set of *logical equivalence classes* of boolean formulas, i.e., a finite subset of a boolean algebra A, as we have being doing in all these chapters. If, and only if, there are no conditions on the events,

A can be assumed to be a free boolean algebra. See Sect. 1.4 for an example where A is not a free boolean algebra.

Chapters 8 and 9 will deal with decidability problems for consistency. Here the input strings for our algorithms are finitely many formulas in Łukasiewicz logic and corresponding rational betting odds. Consistency requires that logically equivalent formulas be assigned equal betting odds. Needless to say, we will have to ensure that this precondition for consistency can be preliminarily checked by a Turing machine, like other preconditions concerning input grammatical correctness, membership of all rational betting odds in the unit interval, and the like.

Historical and bibliographical notes. For background on boolean propositional logic see, e.g., [8].

Lemma 3.5 is due to Horn and Tarski [7, Theorem 1.22].

In the framework of finite boolean algebras, Theorem 3.2 is the formulation of de Finetti's consistency theorem, [2, 3].

The proof of Lemma 3.6, a necessary tool for the proof of Theorem 3.8, rests on the compactness of the Tychonov cube $[0, 1]^A$. In ZF (Zermelo-Fraenkel) set-theory this is equivalent to the "Boolean Prime Ideal" (BPI) theorem, stating that in every boolean algebra A the set $\mathsf{hom}(A)$ is nonempty. The dependence of Theorem 3.8 on this axiom is rarely made explicit in the vast literature on de Finetti's consistency theorem, also known as his "Dutch Book" theorem.

De Finetti proved both directions of the theorem, respectively in [2, pp. 309–312] and [2, p. 313], also see [4]. Still, "converses" of de Finetti's consistency theorem can be found in the literature.

In finance theory, an "arbitrage opportunity" is an investment with an expected positive net cash flow requiring no initial investment. The "Fundamental Theorem of Asset Pricing" states that a financial market has no arbitrage opportunities if and only if, with respect to some probability distribution, the expected value of the asset's future cash flows, discounted at the riskless interest rate, lies between its current bid and ask prices. For details see [1, Example 5.10, pp. 263–264] and [5, 6, 10]. As noted in [9, p. 107], this theorem is just de Finetti's consistency theorem with discounting thrown in. But de Finetti is not usually given credit in the finance literature for having discovered the same result 40 years earlier.

References

1. Boyd, S., & Vandenberghe, L. (2009). *Convex optimization*. Seventh printing with corrections, Cambridge University Press. https://web.stanford.edu/~boyd/cvxbook/bv_cvxbook.pdf
2. de Finetti, B. (1931). Sul significato soggettivo della probabilità. *Fundamenta Mathematicae, 17*, 298–329. Translated into English as: On the Subjective Meaning of Probability. In P. Monari & D. Cocchi (Eds.), Probabilitá e Induzione, Clueb, Bologna, pp. 291–321, (1993). Reprinted in [**4**, pp. 191–222].
3. de Finetti, B. (1937). La prévision: ses lois logiques, ses sources subjectives. *Annales de l'Institut H. Poincaré, 7*, 1–68; English translation by Henry E. Kyburg Jr., as: Foresight: Its Logical Laws, its Subjective Sources. In H. E. Kyburg & H. E. Smokler, Studies in Subjective

Probability, J. Wiley, New York, pp. 93–158, (1964). Second edition published by Krieger, New York, pp. 53–118, (1980). Reprinted in [**4**, pp. 335–400].
4. de Finetti, B. (2006). *Opere Scelte. (Selected Works)* (Vol. 1). Unione Matematica Italiana, Edizioni Cremonese, Firenze.
5. Delbaen, F., & Schachermayer, W. (1994). A general version of the fundamental theorem of asset pricing. *Mathematische Annalen, 300*(1), 463–520.
6. Duffie, D. (1996). *Dynamic asset pricing theory*. Princeton University Press.
7. Horn, A., & Tarski, A. (1948). Measures in Boolean algebras. *Transactions of the American Mathematical Society, 64*, 467–497.
8. Mendelson, E. (1997). *Introduction to mathematical logic* (4th ed.). Chapman & Hall, London.
9. Nau, R. F. (2001). De Finetti was right: Probability does not exist. *Theory and Decision, 51*, 89–124. https://doi.org/10.1023/A:1015525808214
10. Ross, S. A. (1976). Return, risk and arbitrage. In I. Friend & J. Bicksler (Eds.), *Risk and return in finance*. Ballinger.

Chapter 4
Logic-Algebraic Independence and the Product Law

There is a classical probability-free notion of the logic independence of two sets A and B of events, in the sense that no conclusion about an event of B can be drawn knowing the outcome of the events in A, and vice versa. On the other hand, in probability textbooks, events a and b are said to be stochastically independent if the probability of $a \wedge b$ is the product of the probabilities of a and b. What are the relations between these two notions of independence? What is the role of product for logically independent events? This chapter is devoted to the relationships between logic-algebraic independence and the product law for conjunctions of independent events.

4.1 Probability-Free Independence and the Boolean Product Law

Let us consider the following remark by Mark Kac, on page 10 of his 1959 book on Statistical Independence in Probability, Analysis and Number Theory:

> There is, thus, independence in a vague and intuitive sense, and there is "independence" in the narrow but well-defined sense that the rule of product of probabilities is applicable. It was the vague and intuitive notion that provided for a long time the main motivation and driving force behind probability theory. And while an impressive formalism was being created, mathematicians (with very few exceptions) remained aloof because it was not clear to them what the objects were to which the formalism was applicable.

Let g and m be the events "Goldbach's conjecture will be proved next month" and "a new massless particle will be discovered next month". The outcome of g is "independent" of the outcome of m even if we do not assign probabilities to these two events and to their conjunction. One may wonder why the independence of g and m—a logic-algebraic notion that will be made precise in Definition 4.1—should

D. Mundici, *Algebraic Probabilistic Consistency*, Trends in Logic 69,
https://doi.org/10.1007/978-3-031-98337-5_4

preliminarily assign numerical values to g and m. Furthermore, among all possible operations to assess the probability of a conjunction of two "independent" events, what is so special about product?

A well established definition of what Kac called "the vague and intuitive notion of independence" exists in logic and algebra. Theorem 4.3 will show that this classical notion, *combined with de Finetti's consistency notion*, gives to the product operation a fundamental role in the assessment of the probability of the conjunction of logically independent events.

Firstly, in the *uniqueness* part of Theorem 4.3 we will prove that if there exists a consistency preserving continuous operation $\diamond$ for the probability $P(a) \diamond P(b)$ of the conjunction of events belonging to two finite sets of logically independent events, then $\diamond$ coincides with product. The proof is self-contained.

Next, the less elementary part of Theorem 4.3 dealing with the *existence* of a consistency preserving operation, shows that product does preserve the consistency of any arbitrary book of conjunctions of logically independent events, and is thereby characterized by this preservation property.

The gist of Theorem 4.3 is as follows: should Bookie systematically adopt an operation other than product to compute the betting odds of conjunctions of events taken from two consistent books on logically independent events, there are simple cases where Betty can bankrupt him.[1] This will never happen if he adopts product.

Throughout this book "independence" will be understood in the logic-algebraic sense.

Independent Boolean algebras. In combination with de Finetti's consistency notion, the following classical notion of independence will be shown to have a key role in our analysis of logic-algebraic independence:

Definition 4.1 Two subalgebras A, B of a boolean algebra C are said to be *independent* (in C) if $0 \neq a \in A,\ 0 \neq b \in B$ implies $0 \neq a \wedge b$.

Following standard usage, for any function f we write $\mathsf{dom}(f)$ for the domain of definition of f.

Definition 4.2 A quintuple (A, B, C, α, β) is *amenable* if A and B are independent subalgebras of the boolean algebra C, and α and β are $[0, 1]$-valued consistent maps defined on finite sets $\mathsf{dom}(\alpha) \subseteq A$ and $\mathsf{dom}(\beta) \subseteq B$, both containing the unit element 1 of C as a member.[2]

A map $\diamond : [0, 1]^2 \to [0, 1]$ is *(A, B, C, α, β)-consistency preserving* if

(*) $a, a' \in \mathsf{dom}(\alpha),\ b, b' \in \mathsf{dom}(\beta),\quad a \wedge b = a' \wedge b' \Rightarrow \alpha(a) \diamond \beta(b) = \alpha(a') \diamond \beta(b')$,

(**) and the resulting *function*

---

[1] This should not worry real-life bookmakers, who have little inclination to swap their roles with bettors.

[2] With no essential loss of generality, given the assumed consistency of α and β.

$$\alpha \diamond \beta : a \wedge b \in \{p \wedge q \in C \mid p \in \mathsf{dom}(\alpha),\ q \in \mathsf{dom}(\beta)\} \mapsto \alpha(a) \diamond \beta(b) \in [0, 1]$$

is a consistent joint extension of α and β, called the *$\diamond$-book of α and β.*

We say that $\diamond$ *preserves consistency* if it is (A, B, C, α, β)-consistency preserving for any amenable quintuple (A, B, C, α, β).

Theorem 4.3 *Product is the only* consistency preserving $[0, 1]$-*valued continuous operation* $\diamond$ *on* $[0, 1]^2$. *In other words, for any amenable quintuple* (A, B, C, α, β), *condition* (*) *holds when* $\diamond$ *is product, the map* $\alpha \cdot \beta \colon a \wedge b \in \mathsf{dom}(\alpha) \bigwedge\!\!\!\bigwedge \mathsf{dom}(\beta) \mapsto \alpha(a) \cdot \beta(b)$ *is a consistent joint extension of* α *and* β, *and no other continuous map has these properties.*

Here $\mathsf{dom}(\alpha) \bigwedge\!\!\!\bigwedge \mathsf{dom}(\beta)$ is shorthand for $\{a \wedge b \in C \mid a \in \mathsf{dom}(\alpha),\ b \in \mathsf{dom}(\beta)\}$.

4.2 The Product Law Theorem: Proof of Uniqueness

To prove that there is at most one $[0, 1]$-valued consistency preserving continuous operation on $[0, 1]^2$, let $\diamond\colon [0, 1]^2 \to [0, 1]$, be continuous and consistency preserving. In particular, $\diamond$ is (A, B, C, α, β)-consistency preserving for the totality of quintuples (A, B, C, α, β) satisfying the following conditions:

Condition 1. For M, N fixed but otherwise arbitrary integers > 0,

$$A = \{P \times \{1, \ldots, N\} \mid P \subseteq \{1, \ldots, M\}\}, \quad B = \{\{1, \ldots, M\} \times Q \mid Q \subseteq \{1, \ldots, N\}\},$$

and

$$C = \mathsf{pow}(\{1, \ldots, M\} \times \{1, \ldots, N\}). \tag{4.1}$$

One immediately verifies that A and B are independent subalgebras of C, and $A \cup B$ generates C. The map $(d, e) \in \mathsf{at}(A) \times \mathsf{at}(B) \mapsto d \wedge e \in C$ is a bijection of $\mathsf{at}(A) \times \mathsf{at}(B)$ onto $\mathsf{at}(C)$.

Condition 2. The consistent $[0, 1]$-valued maps α and β, are defined as follows:

$$\alpha \text{ is the only state of } A \text{ assigning the value } 1/M \text{ to each atom of } A, \tag{4.2}$$

$$\beta \text{ is the only state of } B \text{ assigning the value } 1/N \text{ to each atom of } B. \tag{4.3}$$

It follows that $1 \in \mathsf{dom}(\alpha) = A$ and $1 \in \mathsf{dom}(\beta) = B$.

By direct inspection, any quintuple (A, B, C, α, β) satisfying the above conditions is amenable. The assumed consistency preservation property of $\diamond$ for any quintuple (A, B, C, α, β) in the above family yields a *consistent* $[0, 1]$-*valued map*

$$\alpha \diamond \beta \colon \mathsf{dom}(\alpha) \bigwedge\!\!\!\bigwedge \mathsf{dom}(\beta) = \{p \wedge q \in C \mid p \in \mathsf{dom}(\alpha),\ q \in \mathsf{dom}(\beta)\} \to [0, 1]$$

that satisfies

$$(\alpha \diamond \beta)(a \wedge b) = \alpha(a) \diamond \beta(b) \text{ for any } a \in A \text{ and } b \in B, \tag{4.4}$$

and is a joint extension of α and β. By (4.1)–(4.3), we can write

$$\mathsf{dom}(\alpha) \cup \mathsf{dom}(\beta) = A \cup B \subseteq \mathsf{dom}(\alpha) \bigwedge\!\!\bigwedge \mathsf{dom}(\beta). \tag{4.5}$$

Theorem 3.2 now yields a state σ of C extending the consistent map $\alpha \diamond \beta$,

$$\sigma \supseteq \alpha \diamond \beta \supseteq \alpha \cup \beta. \tag{4.6}$$

Claim 1: The state σ is uniquely determined by α and β. As a matter of fact, for every atom d of A and e of B, the atom $d \wedge e$ of C satisfies the identities:

$$\sigma(d \wedge e) = \alpha(d) \diamond \beta(e) = \frac{1}{M} \diamond \frac{1}{N} = \frac{1}{M} \cdot \frac{1}{N} \text{ for all integers } M \text{ and } N > 0. \tag{4.7}$$

In view of (4.2)–(4.3), we have only to prove $\frac{1}{M} \diamond \frac{1}{N} = \frac{1}{M} \cdot \frac{1}{N}$. To this purpose, our hypotheses on the admissible quintuple (A, B, C, α, β), and the additivity and extension properties of the state σ yield the following identities:

$$\begin{aligned}
\frac{1}{M} &= \alpha(d) = \sigma(d) = \sigma((d \wedge e) \vee (d \wedge \neg e)) = \sigma(d \wedge e) + \sigma(d \wedge \neg e) \\
&= \alpha(d) \diamond \beta(e) + \sigma(d \wedge \neg e) = \frac{1}{M} \diamond \frac{1}{N} + \sigma(d \wedge \neg e) \text{ by (4.2)-(4.3)} \\
&= \frac{1}{M} \diamond \frac{1}{N} + \sigma(d \wedge \bigvee\{e' \in \mathsf{at}(B) \mid e' \neq e\}), \text{ by Proposition 1.3} \\
&= \frac{1}{M} \diamond \frac{1}{N} + \sigma(\bigvee\{d \wedge e' \mid e' \in \mathsf{at}(B),\ e' \neq e\}), \text{ by distributivity} \\
&= \frac{1}{M} \diamond \frac{1}{N} + \sum\{\sigma(d \wedge e') \mid e' \in \mathsf{at}(B),\ e' \neq e\}, \text{ since } \sigma \text{ is additive} \\
&= \frac{1}{M} \diamond \frac{1}{N} + \sum\left\{\frac{1}{M} \diamond \beta(e') \mid e' \in \mathsf{at}(B),\ e' \neq e\right\}, \text{(4.4)–(4.5)} \\
&= \frac{1}{M} \diamond \frac{1}{N} + (N-1)\left(\frac{1}{M} \diamond \frac{1}{N}\right) = N\left(\frac{1}{M} \diamond \frac{1}{N}\right), \text{ by (4.3).}
\end{aligned}$$

This settles our claim.

More generally, a routine variant of the above calculation yields:

$$\text{For all } a \in A \text{ and } N = 1, 2, \ldots, \ \alpha(a) \diamond \frac{1}{N} = \alpha(a) \cdot \frac{1}{N}. \tag{4.8}$$

Since by hypothesis the $\diamond$ operation is *continuous* we have

$$\alpha(a) \diamond 0 = 0 \quad \text{for all } a \in A. \tag{4.9}$$

Claim 2: For all $a \in A$, $N = 1, 2, \ldots,$ and $l \in \{1, \ldots, N\}$,

$$\alpha(a) \diamond \frac{l}{N} = \alpha(a) \cdot \frac{l}{N}. \tag{4.10}$$

As a matter of fact, let $\{e_1, \ldots, e_l\}$ be a subset of $\mathsf{at}(B)$. Since by hypothesis α and β are consistent (indeed, they are states, respectively of A and B), we can write:

$$\begin{aligned}
\alpha(a) \diamond \frac{l}{N} &= \alpha(a) \diamond (\beta(e_1) + \cdots + \beta(e_l)) = \alpha(a) \diamond \beta(e_1 \vee \cdots \vee e_l) \\
&= \sigma(a \wedge (e_1 \vee \cdots \vee e_l)) = \sigma((a \wedge e_1) \vee \cdots \vee (a \wedge e_l)) \\
&= \sigma(a \wedge e_1) + \cdots + \sigma(a \wedge e_l) = \alpha(a) \diamond \beta(e_1) + \cdots + \alpha(a) \diamond \beta(e_l) \\
&= l\left(\alpha(a) \diamond \frac{1}{N}\right) = l\left(\alpha(a) \cdot \frac{1}{N}\right) = \alpha(a) \cdot \frac{l}{N}, \text{ by (4.8).}
\end{aligned}$$

This settles Claim 2.

From (4.9) we then obtain the following extension of (4.10):

$$\text{For all } a \in A, \quad N = 1, 2, \ldots, \quad \text{and } l \in \{0, \ldots, N\}, \quad \alpha(a) \diamond \frac{l}{N} = \alpha(a) \cdot \frac{l}{N}. \tag{4.11}$$

Let now (A, B, C, α, β) range over all quintuples satisfying conditions 1 and 2 above. Then the set of pairs $(\alpha(a), l/N)$ in (4.11) ranges over the set $([0, 1] \cap \mathbb{Q})^2$ of rationals in the unit square, and we may write

$$x \diamond y = x \cdot y \quad \text{for all } (x, y) \in ([0, 1] \cap \mathbb{Q})^2.$$

From the assumed continuity of $\diamond$ it follows that $\diamond$ coincides with product over $[0, 1]^2$. A fortiori, we have reached the desired conclusion that there do not exist two distinct consistency preserving $[0, 1]$-valued continuous operations $\diamond, \diamond'$ on $[0, 1]^2$. $\square$

Remark 4.4 The continuity of $\diamond$ ensures that negligible errors/perturbations in the assessment of the probabilities $\alpha(x), \beta(y)$ of the events $x \in \mathsf{dom}(\alpha)$, $y \in \mathsf{dom}(\beta)$ have a negligible effect on the values $\alpha(x) \diamond \beta(y)$ assigned by the $\diamond$-book of α and β to the event $x \wedge y$. Dropping the continuity hypothesis, still the consistency preservation property of $\diamond$ ensures that $\diamond$ coincides with product on the dense set of all rational points in $[0, 1]^2$, which is amply sufficient for real-life bookmaking.

4.3 The Product Law Theorem: Stone and Carathéodory

The proof that product is a consistency preserving operation requires more sophistication on the part of the reader than the uniqueness proof of Theorem 4.3. Some exposure with Stone duality and Carathéodory's extension theorem is necessary. For background the reader is referred to the textbooks cited in the historical and bibliographical notes at the end of this chapter.

Definition 4.5 A *maximal ideal* $\mathfrak{m}$ of a boolean algebra A is the kernel of a homomorphism $\eta \in \mathsf{hom}(A)$,

$$\mathfrak{m} = \mathsf{ker}\,(\eta) = \{e \in A \mid \eta(e) = 0\}, \text{ with } \eta \text{ a homomorphism of } A \text{ into } \{0, 1\}.$$

The *Stone space* of A is the set $\boldsymbol{\mu}(A)$ of maximal ideals of A equipped with the (hull-kernel) topology, whose basic closed sets have the form

$$X_a = \{\mathfrak{m} \in \boldsymbol{\mu}(A) \mid a \in \mathfrak{m}\}, \text{ for } a \text{ any element of } A.$$

Any set X_a is also open, whence it is *clopen* (closed-and-open). It follows that any open set of $\boldsymbol{\mu}(A)$ is an arbitrary union of clopen sets of the form X_a. The maximal spectral space $\boldsymbol{\mu}(A)$ is known as a *boolean space*, i.e., a compact Hausdorff space Y whose open sets are precisely the unions of clopen sets of Y. The topological product of two boolean spaces is easily seen to be a boolean space.

Theorem 4.6 (Stone) *For every boolean algebra A, let* $\mathsf{clop}(\boldsymbol{\mu}(A))$ *denote the boolean algebra of clopen sets of the Stone space* $\boldsymbol{\mu}(A)$*, equipped with the distinguished elements* $\emptyset$ *and* $\boldsymbol{\mu}(A)$*, and with the operations of complementation, union, and intersection. Then* $\mathsf{clop}(\boldsymbol{\mu}(A))$ *is isomorphic to A. Every boolean algebra A is isomorphic to the boolean algebra of clopen sets of some boolean space* $Y \neq \emptyset$*:* $A \cong (\mathsf{clop}(Y), \emptyset, Y, \neg, \cup, \cap)$*, with* $\neg$ *denoting the complement operation* $Z \mapsto Y \setminus Z$. □

While Proposition 1.3 identifies every finite boolean algebra with the powerset algebra of a finite set (with the discrete topology), Stone's theorem yields a much more general concrete representation of every boolean algebra A as the algebra of clopen sets in the maximal ideal space $\boldsymbol{\mu}(A)$. When A is finite algebra, $\boldsymbol{\mu}(A)$ can be identified with the set of atoms of A equipped with the discrete topology.

Algebraically independent finite boolean algebras: a construction. Generalizing (1.4), for any (finite or infinite) set $X \neq \emptyset$ we let $\mathsf{pow}(X)$ denote the powerset of X equipped with union, intersection and complement. This is a boolean algebra with $2^{|X|}$ elements, where $|X|$ denotes the cardinality of X. If X is finite, the set $\mathsf{at}(\mathsf{pow}(X))$ of atoms of the powerset boolean algebra $\mathsf{pow}(X)$ is in one-one correspondence with X, and also with the set $\mathsf{hom}(\mathsf{pow}(X))$ of homomorphism of the boolean algebra $\mathsf{pow}(X)$ into the two-element boolean algebra $\{0, 1\}$. By Proposition 1.3, any finite boolean algebra is isomorphic to $\mathsf{pow}(Y)$ for some finite set Y.

For any (finite or infinite) boolean algebra A, the set $\mathsf{hom}(A)$ conveniently replaces the set $\mathsf{at}(\mathsf{pow}(X))$.

As a warmup, the following construction provides a concrete realization of the independence of two subalgebras A, B of a finite boolean algebra C.

Construction 4.7 *Let A and B be finite boolean algebras. By Proposition 1.3 there are finite nonempty sets X and Y such that, without loss of generality,*

$$A = \mathsf{pow}(X), \text{ and } B = \mathsf{pow}(Y). \tag{4.12}$$

We may also assume that X and Y are nonempty disjoint finite sets of nonzero positive integers. So let us arrange the elements of X on the x-axis of the cartesian plane $\mathbb{R}^2$ and similarly arrange the elements of Y on the y-axis. Cylindrification along the y-axis transforms every subset S of X into the subset $S^\uparrow$ of $X \times Y$ given by $S^\uparrow = \{(m, n) \in X \times Y \mid m \in S,\ n \in Y\}$. Symmetrically, cylindrification along the x-axis transforms every subset T of Y into the set $T^\rightarrow = \{(m, n) \in X \times Y \mid m \in X,\ n \in T\}$. Letting S and T range over arbitrary subsets of X and of Y, the cylindrification map $S \mapsto S^\uparrow$ yields a one-one homomorphism (= an embedding) of the boolean algebra $\mathsf{pow}(X)$ into $\mathsf{pow}(X \times Y)$. Symmetrically, the map $T \mapsto T^\rightarrow$ embeds $\mathsf{pow}(Y)$ into $\mathsf{pow}(X \times Y)$. Let $A^\uparrow$ and $B^\rightarrow$ be defined by

$$A^\uparrow = \{S^\uparrow \mid S \subseteq X\}, \quad B^\rightarrow = \{T^\rightarrow \mid T \subseteq Y\}.$$

We claim that $A^\uparrow$ and $B^\rightarrow$ are independent subalgebras of $\mathsf{pow}(X \times Y)$.

As a matter of fact, for any $S \subseteq X$ and $T \subseteq Y$, let us assume $0 \neq S^\uparrow \in A^\uparrow$ and $0 \neq T^\rightarrow \in B^\rightarrow$. This means that for some $m^* \in S$, $S^\uparrow \supseteq \{(m^*, n) \mid n \in Y\} \neq \emptyset$. Likewise, for some $n^* \in Y$, $T^\rightarrow \supseteq \{(m, n^*) \mid m \in X\} \neq \emptyset$. Then the pair (m^*, n^*) is an element of $S^\uparrow \cap T^\rightarrow$, which settles our claim.

Furthermore, the finite set $A^\uparrow \cup B^\rightarrow$ generates $\mathsf{pow}(X \times Y)$, because already the totality of cylindrifications $m^\uparrow$ of points of X and $n^\rightarrow$ of points of Y generate the finite set $\mathsf{pow}(X \times Y)$.

This construction shows that for all finite nonempty sets X, Y there is a bijection of $X \times Y$ onto the set of atoms of the boolean algebra $\mathsf{pow}(X \times Y)$.

Corollary 4.8 *If A and B are independent subalgebras of F and the set $A \cup B$ generates F, then F is isomorphic to the boolean algebra $\mathsf{clop}(\mu(A) \times \mu(B))$, where $\mu(A) \times \mu(B)$ is equipped with the product topology. Thus the Stone space of F is homeomorphic to the product of the Stone spaces of A and B, $\mu(F) \cong \mu(A) \times \mu(B)$.*

Proof Combine 4.7 with Stone's theorem 4.6. □

F is known as the *free product* of A and B, and is variously denoted in the literature as $A \amalg B$, $A \oplus B$, $A \otimes B$, ... We will adopt the notation $A \amalg B$.

Definition 4.9 For any nonempty set Ω, by a *σ-algebra in Ω* we mean a subset of the powerset $\mathsf{pow}(\Omega)$ containing Ω and closed under complementation and countable unions. A pair $(\Omega, \mathcal{A})$ consisting of a nonempty set Ω and a σ-algebra $\mathcal{A} \subseteq \mathsf{pow}(\Omega)$ is called a *measurable space*. A function $\nu\colon \mathcal{A} \to \mathbb{R} \cup \{\infty\}$ with $\nu(\emptyset) = 0$ is said to be *finite* if its range does not contain ∞. If there is a sequence $\Omega_1, \Omega_2, \ldots$ of subsets of Ω such that $\Omega = \bigcup_n \Omega_n$ and $\nu(\Omega_n) < \infty$ for all n, ν is *σ-finite*.

We say that ν is a *measure* on the measurable space $\mathcal{A}$ if ν has the following *σ-additivity property*: For any sequence of pairwise disjoint subsets Ω_i of Ω, $\nu(\bigcup_n \Omega_n) = \sum_n \nu(\Omega_n)$. If, in addition, $\nu(\Omega) = 1$, ν is a *probability measure.* A triple $(\Omega, \mathcal{A}, \nu)$ is called a *measure (resp., a probability) space* if $(\Omega, \mathcal{A})$ is a measurable space and ν is a measure (resp., a probability measure) on $\mathcal{A}$. $(\Omega, \mathcal{A}, \nu)$ is *σ-finite* if so is ν. Trivially, every probability measure is σ-finite.

When Ω is a topological space the σ-algebra $\mathcal{B}(\Omega)$ generated by all open sets of Ω is called the *Borel* σ-algebra on Ω, and each element of $\mathcal{B}(\Omega)$ is a *Borel set.* Every measure ν on $\mathcal{B}(\Omega)$ is said to be a *Borel measure.* The pair $(\Omega, \mathcal{B}(\Omega))$ is a *Borel measurable space.* The triple $(\Omega, \mathcal{B}(\Omega), \nu)$ is said to be a *Borel measure space.* For any Borel measurable spaces $(\Omega_1, \mathcal{B}(\Omega_1))$ and $(\Omega_2, \mathcal{B}(\Omega_2))$ we denote by $\mathcal{B}(\Omega_1 \times \Omega_2)$ the σ-algebra of subsets of $\Omega_1 \times \Omega_2$ generated by all sets of the form $X \times \Omega_2$ and $\Omega_1 \times Y$, where $X \in \mathcal{B}(\Omega_1)$ and $Y \in \mathcal{B}(\Omega_2)$. $\mathcal{B}(\Omega_1 \times \Omega_2)$ is a Borel measurable space.

Theorem 4.10 (Carathéodory) *Let X be a boolean space. Then every state of the boolean algebra* $\mathsf{clop}(X)$ *uniquely extends to a probability measure on the σ-algebra $\mathcal{B}(X)$ of Borel sets of X.* □

Corollary 4.11 *Let X be a boolean space and β a* $[0, 1]$*-valued map defined on a finite subset of the boolean algebra* $\mathsf{clop}(X)$. *Then β is consistent if and only if it can be extended to a regular Borel probability measure on X.*

Proof Combine Theorem 4.10 with the part of de Finetti's consistency Theorem 3.8 showing that consistency implies extendability to a state. □

We will also use the following result:

Theorem 4.12 (σ-finite product measure) *Let $(\Omega, \mathcal{B}(\Omega), \mu)$ and $(\Omega', \mathcal{B}(\Omega'), \mu')$ be σ-finite Borel measure spaces. Then there is a unique σ-finite measure ν on the Borel σ-algebra $\mathcal{B}(\Omega \times \Omega')$ satisfying $\nu(E \times E') = \mu(E) \cdot \mu'(E')$ for all $E \in \mathcal{B}(\Omega)$ and $E' \in \mathcal{B}(\Omega')$. We say that ν is the* product measure of *μ and μ'.* □

4.4 The Product Law Theorem: Conclusion of the Proof

We have already proved the uniqueness part of Theorem 4.3, stating that there is *at most one* consistency preserving continuous operation $\diamond\colon\ [0, 1]^2 \to [0, 1]$. Actually, we have also proved that, if such operation exists it coincides with product. To conclude the proof of Theorem 4.3 we must show that the product operation does have the consistency preservation property.

Let the boolean algebras A, B, C and the consistent maps α, β be as in Definition 4.2. In view of Stone's Theorem 4.6, without loss of generality we may argue under the assumption

$$A = \mathsf{clop}(X) \quad \text{and} \quad B = \mathsf{clop}(Y), \text{ for some boolean spaces } X \text{ and } Y.$$

Since A and B are independent boolean subalgebras of C, they are also independent subalgebras of the (free product) boolean algebra $A \amalg B$ of C generated by the set $A \cup B$. Without loss of generality, we may henceforth assume $C = A \amalg B$. Furthermore, by Corollary 4.8, we may argue assuming $A \amalg B = \mathsf{clop}(X \times Y)$, where the boolean space $X \times Y$ is the cartesian product of X and Y equipped with the product topology.

From the ($\Rightarrow$)-direction of Theorem 3.8 there exist states $\bar{\alpha}$ of A and and $\bar{\beta}$ of B, respectively extending the consistent maps α and β. Theorem 4.10 yields Borel probability measures $\bar{\bar{\alpha}}$ and $\bar{\bar{\beta}}$ over the Borel σ-algebras $\mathcal{B}(X)$ and $\mathcal{B}(Y)$ generated by the clopen sets of the spaces X and Y, respectively extending $\bar{\alpha}$ and $\bar{\beta}$. In symbols,

$$\bar{\alpha}(a) = \bar{\bar{\alpha}}(a) \text{ for all } a \in \mathsf{clop}(X) \text{ and } \bar{\beta}(b) = \bar{\bar{\beta}}(b) \text{ for all } b \in \mathsf{clop}(Y).$$

Theorem 4.12 applied to the probability spaces $(X, \mathcal{B}(X), \bar{\bar{\alpha}})$ and $(Y, \mathcal{B}(Y), \bar{\bar{\beta}})$ now yields a unique probability measure $\bar{\bar{\mu}}$ on $\mathcal{B}(X \times Y)$ satisfying

$$\bar{\bar{\mu}}(U \times V) = \bar{\bar{\alpha}}(U) \cdot \bar{\bar{\beta}}(V) \text{ for all } U \in \mathcal{B}(X) \text{ and } V \in \mathcal{B}(Y).$$

Let the function $\bar{\mu}$ be the restriction of $\bar{\bar{\mu}}$ to $A \amalg B = \mathsf{clop}(X \times Y)$. Since $\bar{\mu}(c) = \bar{\bar{\mu}}(c)$ for all $c \in A \amalg B$, then $\bar{\mu}$ is a state of the boolean algebra $A \amalg B$. For all $a, a' \in A$ and $b, b' \in B$ we have

$$\text{if } a \wedge b = a' \wedge b' \text{ then } \bar{\mu}(a \wedge b) = \bar{\alpha}(a) \cdot \bar{\beta}(b) = \bar{\alpha}(a') \cdot \bar{\beta}(b'). \tag{4.13}$$

Let the set $A \wedge B \subseteq A \amalg B$ be defined by

$$A \wedge B = \{a \wedge b \mid a \in A \text{ and } b \in B\}.$$

By (4.13) the restriction of $\bar{\mu}$ to $A \wedge B$ yields a *function* $\mu \colon A \wedge B \to [0, 1]$. Since the states $\bar{\alpha}$ and $\bar{\beta}$ respectively extend the consistent maps α and β, we can now immediately infer that μ is a joint extension of α and β, and μ is the product book of α and β on $A \wedge B$, $\mu = \alpha \cdot \beta$. Since μ is the restriction to $A \wedge B$ of a state of the boolean algebra $A \amalg B$, from the ($\Leftarrow$)-direction of Theorem 3.8 it follows that μ is a consistent map.

The proof of Theorem 4.3 is now complete. □

In conclusion:

Product is the only consistency preserving continuous operation for conjunctions of yes-no events taken from two consistent books on independent sets of events.

4.5 Finite Versus σ-Additive Probability on Boolean Spaces

On page 60 of his 1998 Fubini lectures [9] Rota writes:

> Pointless probability deals with an abstract Boolean σ-algebra Π and with a real-valued function defined on Π which imitates the definition of probability. The problem is to define an algebra of random variables. The setup is not as artificial as it may appear. Among probabilists, mention of sample points in an argument has always been bad form. A fully probabilistic argument must be pointless.

In Kolmogorov's probability theory, "events" form a σ-*algebra*, i.e., a boolean algebra closed under countable unions, which is thought of by him as being more fundamental than the space Ω of sample points, alias, "elementary events". On page 61 of his paper [4] Kolmogorov writes:

> ...the notion of an elementary event is an artificial superstructure imposed on the concrete notion of an event. In reality, events are not composed of elementary events, but elementary events originate in the dismemberment of composite events.

Boolean spaces of sample points and yes-no boolean events are reconciled by Stone representation, showing that the space X of elementary events with its topology corresponds to its dual given by the set of events equipped with their boolean algebraic structure A. From the viewpoint of algebraic probability, the duality between A and X is further strengthened by Theorem 4.10, yielding an affine isomorphism between the regular Borel probability measures on X and the states of A. Combining this theorem with the consistency Theorems 3.8, Theorem 4.14 will highlight the correspondence between boolean algebras B with a distinguished state α and Kolmogorov triples $(\Omega, \mathcal{B}(\Omega), P)$, for Ω any arbitrary totally disconnected compact Hausdorff space.

The necessary preliminaries are as follows.

Definition 4.13 Let us agree to say that an *enriched boolean algebra* is a pair (B, α) where B is a boolean algebra and α is a state (i.e., a finitely consistent map) of B. Two enriched boolean algebras (B, α) and (B', α') are *isomorphic* if there is an isomorphism $\phi\colon B \to B'$ such that $\alpha = \alpha'(\phi)$. In symbols, $\phi\colon (B, \alpha) \cong (B', \alpha')$.

Given a homeomorphism $\psi\colon \Omega \cong \Omega'$, the isomorphism of two Kolmogorov triples $(\Omega, \mathcal{B}(\Omega), P)$ and $(\Omega', \mathcal{B}(\Omega'), P')$ is defined in a similar way.

For X a boolean space, let $\overline{X}$ be the boolean algebra of clopen sets of X. By Theorem 4.6, up to isomorphism, $\overline{X}$ is the most general possible boolean algebra. For every boolean algebra B let $\overline{B}$ denote its Stone space. Without danger of confusion we may write

$$\overline{\overline{B}} = B \quad \text{and} \quad \overline{\overline{X}} = X.$$

Theorem 4.14 *Let the map ϕ transform any boolean algebra B enriched by a finitely consistent map α into the Kolmogorov triple $(\overline{B}, \mathcal{B}(\overline{B}), \alpha')$ with α' the regular Borel probability measure on $\overline{B}$ corresponding to α by Carathéodory extension Theorem 4.10.*

Conversely, let the map ψ transform every Kolmogorov triple $(\Omega, \mathcal{B}(\Omega), P)$, (with Ω a boolean space and P a probability measure on $\mathcal{B}(\Omega)$), into the boolean algebra $\overline{\Omega}$ enriched by the finitely consistent map $P \upharpoonright \overline{\Omega}$ = the restriction P to $\overline{\Omega}$.

Then $\psi(\phi(B, \alpha)) \cong (B, \alpha)$ and $\phi(\psi(\Omega, \mathcal{B}(\Omega), P)) \cong (\Omega, \mathcal{B}(\Omega), P)$.

Proof By Corollary 3.10, the states of any boolean algebra are precisely its finitely consistent maps. A $[0, 1]$-valued map β defined on a finite subset of the Stone dual $\overline{X}$ of a boolean space X is consistent in $\overline{X}$ if and only if it can be extended to a regular Borel probability measure P_β on X. As a matter of fact,

— If β is consistent in $\overline{X}$ then by the consistency Theorem 3.8, β can be extended to a finitely consistent map σ_β of $\overline{X}$. By Theorem 4.10, σ_β is extendable to a regular Borel probability measure P_β on X.

— Conversely, if β is extendable to a probability measure Q on the space $\mathcal{B}(X)$ of Borel sets of X then, letting $Q \upharpoonright \overline{X}$ denote the restriction of Q to $\overline{X}$, β is trivially extendable to $Q \upharpoonright \overline{X}$. By the converse direction of the Theorem 3.8, β is consistent in the boolean algebra $\overline{X}$.

Identifying B and $\overline{\overline{B}}$, we have just proved that for any boolean algebra B the consistent maps of B coincide with the finite restrictions of regular Borel probability measures on the Stone space $\overline{B}$ of B. □

Remark 4.15 By Theorem 4.14, Kolmogorov's countably additive probability on boolean spaces lives in symbiosis with de Finetti's boolean algebraic finitely additive probability. This said, totally disconnected compact Hausdorff spaces and $\{0, 1\}$-events are inadequate to recover Kolmogorov's probability, e.g., *on compact spaces*. To this purpose, in the second part of the book Theorem 4.14 will be extended to continuous $[0, 1]$-events and their states. MV-algebras stand to Łukasiewicz logic as boolean algebras stand to boolean logic. Their central role in the algebraic theory of continuous events is highlighted by the characterization of Łukasiewicz logic in the next chapter, and by several further results on the role of consistency in the algebraic probabilistic theory of continuous events. Boolean algebras are just idempotent MV-algebras.

Historical and bibliographical notes. Kac's quotation at the beginning of this chapter can be found in [2].

For details on boolean algebraic independence see [5, Chapter 4, 11.3] or [10, §13]. As noted at the outset of this chapter, the independence of two sets of events A and

B means that no conclusion about an event of B can be drawn knowing the outcome of the events in A, and vice versa. To be precise, if the formula α codes $a \in A$, β codes $b \in B$ and $\alpha \rightarrow \beta$ is a tautology, then either β itself is a tautology, or else $\neg\alpha$ is a tautology. This is easily seen to be equivalent to the algebraic independence of A and B, as defined Definition in 4.1. Equivalently, all elements of A can be coded by boolean formulas none of whose variables occur in the formulas coding the events in B. This is a special case of Craig interpolation theorem, [6, 1.45(b), p. 33]. For boolean events, logic thus provides an alternative intuitive formulation of independence.

Our standard reference for topology is [1].
A proof of Corollary 4.8 can be found in [5, (1), p. 161]. For a proof of Theorem 4.6 see, e.g., [5, 7.1–7.3]. Corollary 4.8 is proved in [5, (1), p. 161], using a different but equivalent notation.

Theorem 4.10 is proved in [11, Corollary 1] and [12, Corollary 2.1.4]. A proof can also be obtained from [3, Theorem 1.41, p. 18], upon noting that if a countable union of pairwise disjoint clopen subsets $C_1, C_2, \ldots$ of X is clopen then only finitely many C_i are nonempty.

For a proof of Theorem 4.12 see [3, Theorem 14.14, p. 277] or [13, Proposition 1.7.11].

Theorem 4.3 was first proved in [7] and [8].

References

1. Engelking, R. (1989). General topology. Revised and completed edition. *Sigma series in pure mathematics* (Vol. 6). Herdermann, Berlin.
2. Kac, M. (1959). Statistical independence in probability, analysis and number theory. In *The Carus mathematical monographs* (Vol. 12). The Mathematical Association of America. Wiley.
3. Klenke, A. (2013). *Probability theory, a comprehensive course* (2nd ed.). Universitext, Springer.
4. Kolmogorov, A. N. (1995). Complete metric Boolean algebras. *Philosophical Studies: An International Journal for Philosophy in the Analytic Tradition, 77*(1), 57–66.
5. Koppelberg, S. (1989). General theory of Boolean algebras. In *Handbook of Boolean algebras* (Vol. 1). Edited by J. D. Monk with the cooperation of R. Bonnet, Elsevier, Amsterdam.
6. Mendelson, E. (1997). *Introduction to mathematical logic* (4th ed.). Chapman & Hall.
7. Mundici, D. (2017). Coherence of de Finetti coherence. *Synthese, 194*, 4055–4063.
8. Mundici, D. (2019). De Finetti coherence and the product law for independent events. *Synthese, 196*, 265–271.
9. Rota, G-C. (2001). Twelve problems in probability no one likes to bring up. The Fubini Lectures, Turin, 3–5 June, 1998. In H. Crapo & D. Senato (Eds.), *Algebraic combinatorics and computer science, a tribute to Gian-Carlo Rota* (pp. 57–93). Springer.
10. Sikorski, R. (1960). *Boolean algebras* (3rd ed.). Ergebnisse der Mathematik und ihrer Grenzgebiete. (Vol. 25). Springer.
11. Tao, T. 245B, notes 0a. An alternate approach to the Carathéodory extension theorem. https://terrytao.wordpress.com/2009/01/03/
12. Tao, T. (2010). An epsilon of room I: Real analysis. In *Graduate studies in mathematics* (Vol. 117). American Mathematical Society. https://terrytao.wordpress.com/2009/01/03/
13. Tao, T. (2011). *An introduction to measure theory, graduate studies in mathematics* (Vol. 126). American Mathematical Society.

Part II
From {0, 1}-events to [0, 1]-events

In Part II of the book all results proved in Part I will be extended from {0, 1}-valued to [0, 1]-valued events. Just as boolean formulas code {0, 1}-events/statements, formulas in Łukasiewicz infinite-valued propositional logic Ł_∞ more generally code [0, 1]-valued events/propositions/functions.

One may wonder why, in the profusion of non-classical extensions of boolean logic, Ł_∞ has been singled out to deal with algebraic probabilistic consistency for continuous events. A main reason is that Łukasiewicz logic is *the only logic* generated by a continuous [0, 1]-valued operation defined on $[0, 1]^2$ that meets the following two bare minimum conditions satisfied by what we usually understand as a continuous "implication": For all $x, y, z \in [0, 1]$,

$$x \to y = 1 \text{ iff } x \leq y \tag{*}$$

and

$$x \to (y \to z) = y \to (x \to z). \tag{**}$$

The reader may notice that boolean logic is the only logic generated by {0, 1}-valued (automatically continuous) implication $\to$ satisfying Condition (*), whence (trivially) also satisfying Condition (**). It is a remarkable fact that Łukasiewicz logic is characterized by the same conditions, once {0, 1}-events/propositions are generalized to [0, 1]-valued events/propositions. The long elementary proof of this fact is given in Chapter 5.

In Chapter 6 MV-algebraic states are defined as the natural generalization of boolean states, i.e., normalized [0, 1]-valued maps that are additive on disjunctions of incompatible events. With MV-algebraic homomorphisms in place of boolean homomorphisms, also the definition of probabilistic consistency naturally extends to Łukasiewicz logic and MV-algebras. Theorem 6.3 states that the probabilistic consistency of a [0, 1]-valued map β on a finite subset of an MV-algebra A is equivalent to the extendability of β to a state of A. In particular, β is logically consistent if and only if it is extendable to a homomorphism of A into the standard MV-algebra [0, 1].

As a key step for the proof of Theorem 6.3, the basic properties of states in lattice-ordered abelian groups with a distinguished strong order unit (unital ℓ-groups, for short) are transferred to MV-algebras via the categorical equivalence Γ. In Corollary 6.17 we will extend to MV-algebras the solution of Boole's problem of determining the possible probabilities of a continuous event h knowing the probabilities of other continuous events on which h depends.

In Theorem 7.18 the product operation on $[0, 1]^2$ is characterized as the only consistency preserving continuous operation in a book of conjunctions of events taken from any two consistent books of algebraically independent continuous events.

Generalizing Theorem 4.14, in Theorem 7.20 any Kolmogorov's probability triple $(\Omega, \mathcal{B}(\Omega), P)$ on a compact Hausdorff space Ω is shown to arise from some pair (A, σ), with A an MV-algebra enriched by a state σ. Ω is a homeomorphic copy of the maximal spectral space of A, $\mathcal{B}(\Omega)$ is the sigma-algebra of Borel sets of Ω, and P is the regular Borel probability measure on Ω corresponding to σ by Theorem 7.13. The latter theorem yields an affine isomorphism between the state space of any MV-algebra A and the space of regular Borel probability measures on the maximal spectral space of A. While de Finetti's consistency notion is generally considered a cornerstone of finitely additive probability, Theorem 7.20 shows that MV-algebraic states are symbiotic with Kolmogorov countably additive probability.

The logic-algorithmic machinery of Łukasiewicz logic and the geometric-polyhedral structural properties of MV-algebras will be then applied to the recognition problem for consistent and inconsistent books of betting odds. More precisely, Chapter 8 is devoted to the construction of a Turing machine which, on any input consisting of a finite set of pairs of Łukasiewicz formulas ψ_j and rational (betting odds) $r_j \in [0, 1]$, decides if the book described by this set is consistent. In Chapter 9 this decidability result is generalized to any input $((\psi_1, r_1), \ldots, (\psi_m, r_m); \theta)$,with θ a Łukasiewicz formula conditioning the set of possible worlds where the ψ_j take their values. Algebraically speaking, we prove the uniform decidability of the consistency problem for all *finitely presented* MV-algebras. It is of interest to note that for any such MV-algebra A, probabilistic consistency is also equivalent to extendability to a state of A that satisfies the stronger condition of being a convex combination of homomorphisms of A into $[0, 1]$.

As we will see, all the main results in Part II depend on a variety of fundamental theorems in polyhedral topology, measure theory, functional analysis, ordered group theory, and computation theory. Lack of space has made it impossible to provide self-contained proofs of all these preparatory results. At the end of each chapter detailed bibliographical references are given to readers interested in the proofs of these results. On the other hand, as mentioned in the Introduction, for every unproved lemma/proposition/theorem/corollary about Łukasiewicz logic and MV-algebras needed in this book, the reader is addressed without exception to the author's joint book with R. Cignoli and I.M.L. D'Ottaviano, "Algebraic Foundations of Many-valued Reasoning", Trends in Logic, vol. 7. Kluwer, 2000, henceforth denoted "TheBook".

Reverting to boolean algebras and to the fundamental notion of independence, Chapter 10 contains a self-contained proof of de Finetti's exchangeability/representation theorem, (also known as de Finetti's theorem) a cornerstone of statistical inference.

In the final chapter we list a small number of problems.

All MV-algebras in this book are *proper*, i.e., $0 \neq 1$. Every ideal J of an MV-algebra A is *proper*, i.e., $J \neq A$. As in the first part, all topological spaces have the Hausdorff separation property.

Chapter 5
Continuous [0, 1]-Valued Events and Their Algebras

While boolean logic deals with the syntax and the semantics of formulas coding $\{0, 1\}$-valued events/propositions, most random variables in everyday life take a continuum of values.

As a consequence, our measurements/assessments of these variables are never infinitely precise. That is why, e.g., any measurement in physics specifies a real number together with an error interval. Yet, relations between real-valued quantities, notably physical laws, are not formulated as relations between intervals. This is made possible whenever small errors in the measurement of the basic observables have a small effect on the assessment of all derived observables. For any bounded observable one may rescale the unit of measurement in such a way that the result becomes a dimensionless number lying in the interval $[0, 1]$. Once a $[0, 1]$-valued logic L is chosen to deal with $[0, 1]$-valued observables, derived observables can be formalized in L by applying the connectives of L to the formulas coding the basic observables. The latter may be regarded as the "propositional variables" of L. The assumption that small errors in the measurement of the basic observables do not propagate into enormously large errors in the evaluation of the derived observables is taken care of by assuming that every connective codes a continuous operation on the set $[0, 1]$ of truth-values.

5.1 Two Bare Minimum Conditions for a continuous "Implication" $\rightarrow_L$

A well known property of boolean logic, known as "functional completeness", ensures that all n-variable boolean functions are obtainable from the identity functions via the boolean operations $\neg, \wedge, \vee$. By contrast, a trivial counting argument shows that no $[0, 1]$-valued logic L can be functionally complete. Therefore, the definition of any such L will critically depend on our choice of the set of connectives of L.

D. Mundici, *Algebraic Probabilistic Consistency*, Trends in Logic 69,
https://doi.org/10.1007/978-3-031-98337-5_5

No matter our definition of a "logic" L, there must be a binary "consequence" relation $\vdash_L$ specifying L. For any two formulas P, Q of L let us read $P \vdash_L Q$ as "L infers Q from P". A main property of $\vdash_L$ traditionally hinges upon a binary connective $\rightarrow_L$ as follows:

For any two L-formulas A and $(A \rightarrow_L B)$ the logic L infers B.

for short, $A,\ A \rightarrow_L B \ \vdash_L \ B$ (i.e., "modus ponens" holds in L). While $\rightarrow_L$ is mostly read as the verbs "infers" or "implies", it is thought of as an "implication" operation. For our present problem, $\rightarrow_L$ must be a continuous [0, 1]-valued binary operation defined on the set $[0, 1]^2$ of pairs of truth-values. In the boolean case, continuity trivially follows from the discreteness of the topology on the two-element set $\{0, 1\}$.

Since we are looking for an implication connective $\rightarrow_L$ in a logic L for [0, 1]-valued continuous events, we cannot forget that the topological space [0, 1] of truth-values is also a totally ordered set with a largest and a smallest element. If $\rightarrow_L$ is to be reminiscent of what is to be understood by a [0, 1]-valued "implication" defined on $[0, 1]^2$, then for any two truth-values $x, y \in [0, 1]$, $x \rightarrow_L y$ equals 1 precisely when $x \leq y$. This is condition (*) above. In boolean logic this condition characterizes implication, and immediately implies $x \rightarrow_L (y \rightarrow_L z) = y \rightarrow_L (x \rightarrow_L z)$, which amounts to (**).

In Theorem 5.6 we will see that continuity together with conditions (*)-(**) characterize Łukasiewicz infinite-valued calculus $Ł_\infty$.

5.2 First Consequences of the Two Bare Minimum Conditions

In this section we draw several immediate consequences of conditions (*)-(**). The reader is advised to sit down before undertaking this task. The following tedious saga of identities, inequalities, implications will provide the necessary link between the two parts of the book, from boolean to Łukasiewicz logic. Conditions (*)-(**) will be mostly used without explicit mention. The reader will be able to discern which one of them is being applied. Following standard usage, the symbol $\Rightarrow$ is shorthand for "implies", and $\Leftrightarrow$ stands for "iff". The variables $x, y, z, \ldots$ range over $[0, 1]$.

As a warmup, let us prove

$$x \leq y \ \Rightarrow \ y \rightarrow z \leq x \rightarrow z. \tag{5.1}$$

We first write $1 = (y \rightarrow z) \rightarrow (y \rightarrow z) = y \rightarrow ((y \rightarrow z) \rightarrow z)$, whence $y \leq ((y \rightarrow z) \rightarrow z)$. From the hypothesis $x \leq y$ we get $x \leq (y \rightarrow z) \rightarrow z$, whence $1 = x \rightarrow ((y \rightarrow z) \rightarrow z) = (y \rightarrow z) \rightarrow (x \rightarrow z)$, and hence $y \rightarrow z \leq x \rightarrow z$, as desired to conclude the proof of (5.1). The implication

$$x \leq y \quad \Rightarrow \quad \neg x \; \geq \; \neg y \tag{5.2}$$

is a particular case of (5.1), because $x \leq y$ implies $y \to 0 \leq x \to 0$, i.e., $\neg y \leq \neg x$.

To prove the inequality

$$(x \to y) \to y \; \geq \; \max(x, y) \tag{5.3}$$

we argue as follows:

On the one hand, from $1 = (x \to y) \to (x \to y) = x \to ((x \to y) \to y)$ we get

$$x < (x \to y) \to y.$$

On the other hand, from $y \to ((x \to y) \to y) = (x \to y) \to (y \to y) = 1$ we get

$$y \leq (x \to y) \to y,$$

which settles (5.3).

Derived operations and consequences of the continuity of the $\to$ operation. Let the derived $[0, 1]$-valued operations $\neg, \oplus, \odot, \ominus$ be defined by stipulating that, for all $x, y, z \in [0, 1]$

$$\neg x = x \to 0, \quad x \oplus y = \neg x \to y, \quad x \odot y = \neg(x \to \neg y), \quad x \ominus y = x \odot \neg y. \tag{5.4}$$

Since the operation $\to$ is continuous, then so are $\neg$, $\oplus$, $\odot$ and $\ominus$. In the remainder of the proof we will make repeated use of the continuity of these operations.

We begin by proving the identity

$$\neg\neg x = x. \tag{5.5}$$

As a first step, from (5.3) we obtain $x \leq (x \to 0) \to 0 = \neg\neg x$, whence by (5.2), $\neg x \geq \neg\neg\neg x$. Conversely, from $\neg x \to \neg\neg\neg x = \neg x \to (((x \to 0) \to 0) \to 0) = ((x \to 0) \to 0) \to (\neg x \to 0) = ((x \to 0) \to 0) \to ((x \to 0) \to 0) = 1$ we get $\neg x \leq \neg\neg\neg x$, whence

$$\neg x = \neg\neg\neg x. \tag{5.6}$$

As a second step, from $1 = x \to 1 = x \to (y \to y) = y \to (x \to y)$ we obtain $x \to y \; \geq \; y$. Conversely, from $1 = (1 \to y) \to (1 \to y) = 1 \to ((1 \to y) \to y)$ repeated applications of (*) yield $1 \to y \leq y$. As a consequence,

$$y = 1 \to y. \tag{5.7}$$

In particular, $\neg 1 = 1 \to 0 = 0$. From $\neg 0 = 0 \to 0 = 1$ it follows that the range of the continuous function $\neg : z \mapsto \neg z$ coincides with $[0, 1]$. As a consequence, for every $x \in [0, 1]$ there is an element $a_x \in [0, 1]$ such that $x = \neg a_x$. Recalling now (5.6) we can write $\neg\neg x = \neg\neg\neg a_x = \neg a_x = x$, as required to complete the proof of (5.5).

From (5.4)–(5.5) we immediately obtain

$$x \to y = \neg x \oplus y. \tag{5.8}$$

Our next goal is to prove the following (contrapositive) property of the $\to$ operation:

$$x \to y = \neg y \to \neg x. \tag{5.9}$$

To this purpose, by (5.5) we can write $(x \to y) \to (\neg y \to \neg x) = (x \to y) \to (\neg y \to (x \to 0)) = (x \to y) \to (x \to (\neg y \to 0)) = (x \to y) \to (x \to \neg\neg y) = (x \to y) \to (x \to y) = 1$. As a consequence, $x \to y \leq \neg y \to \neg x$. The converse inequality similarly follows from (5.5). This completes the proof of (5.9).

Combining (5.5) and (5.9) we obtain the identities $x \oplus y = \neg x \to y = \neg y \to \neg\neg x = \neg y \to x = y \oplus x$, proving that $\oplus$ is commutative:

$$x \oplus y = y \oplus x. \tag{5.10}$$

To prove that $\oplus$ is associative,

$$x \oplus (y \oplus z) = (x \oplus y) \oplus z, \tag{5.11}$$

by (5.5)–(5.10) we may write $(x \oplus y) \oplus z = \neg(\neg x \to y) \to z = \neg z \to \neg\neg(\neg x \to y) = \neg z \to (\neg x \to y) = \neg x \to (\neg z \to y) = x \oplus (z \oplus y) = x \oplus (y \oplus z)$.

Next to prove the equivalences

$$x \odot y \leq z \;\Leftrightarrow\; x \leq y \to z \;\Leftrightarrow\; y \leq x \to z \;\Leftrightarrow\; y \odot x \leq z \tag{5.12}$$

we argue as follows: Combining (5.1) and (5.5)–(5.9) we have: $x \odot y \leq z \Leftrightarrow \neg(x \to \neg y) \leq z \Leftrightarrow x \to \neg y \geq \neg z \Leftrightarrow \neg z \to (x \to \neg y) = 1 \Leftrightarrow x \to (\neg z \to \neg y) = 1 \Leftrightarrow x \to (y \to z) = 1 \Leftrightarrow x \leq (y \to z)$. Since $x \to (y \to z) = 1 \Leftrightarrow y \to (x \to z) = 1$, interchanging the roles of x and y we obtain the desired conclusion (5.12).

From (5.12) it follows that $x \odot y$ and $y \odot x$ have the same upper bounds in $[0, 1]$, and hence also $\odot$ is commutative:

$$x \odot y = y \odot x. \tag{5.13}$$

The associativity property

$$x \odot (y \odot z) = (x \odot y) \odot z \tag{5.14}$$

now follows from (5.5)–(5.9) and (5.13) by writing $x \odot (y \odot z) = \neg(x \to \neg(y \odot z)) = \neg(x \to \neg\neg(y \to \neg z)) = \neg(x \to (y \to \neg z)) = \neg(x \to (z \to \neg y)) = \neg(z \to (x \to \neg y)) = \neg(z \to \neg\neg(x \to \neg y)) = \neg(z \to \neg(x \odot y)) = z \odot (x \odot y) = (x \odot y) \odot z$.

To prove the monotonicity of the operation $\to$ we first use (5.13) to obtain the identity $x \odot (x \to y) = (x \to y) \odot x$. From (5.12) we have $(x \to y) \odot x \leq y \Leftrightarrow x \to y \leq x \to y$. As a consequence, $x \odot (x \to y) \leq y$. Combining this latter inequality with (5.13) we can write $(x \to y) \odot x = x \odot (x \to y) \leq y$, whence $y \leq z \Rightarrow x \odot (x \to y) \leq z$. Likewise, by (5.13) and (5.12), $x \odot (x \to y) \leq z \Leftrightarrow (x \to y) \odot x \leq z \Leftrightarrow x \to y \leq x \to z$. We have thus proved that $\to$ is monotonic:

$$y \leq z \Rightarrow x \to y \leq x \to z. \tag{5.15}$$

Using (5.15) and (5.5)–(5.9) we immediately obtain the monotonicity of $\odot$:

$$y \leq z \Rightarrow x \odot y \leq x \odot z. \tag{5.16}$$

From (5.8) and (5.15) we also obtain the monotonicity of $\oplus$:

$$y \leq z \Rightarrow x \oplus y \leq x \oplus z. \tag{5.17}$$

We now characterize the $\to$ operation as follows:

$$x \to y = \max\{t \mid x \odot t \leq y\}. \tag{5.18}$$

To this purpose, let the set $W_{x,y} \subseteq [0, 1]$ be defined by

$$W_{x,y} = \{t \in [0, 1] \mid x \odot t \leq y\}.$$

Since $\odot$ is continuous and the interval $[0, 1]$ is compact then $\sup W_{x,y} = \max W_{x,y}$. Since $x \odot 0 = 0$, from (5.16) it follows that $W_{x,y}$ is a nonempty interval in $[0, 1]$. For any $z \in [0, 1]$, an application of (5.12) yields the equivalences

$$\max W_{x,y} \geq z \Leftrightarrow z \in W_{x,y} \Leftrightarrow x \odot z \leq y \Leftrightarrow x \to y \geq z,$$

as required to complete the proof of (5.18).

Using (5.18) and the continuity of $\odot$, our next goal is to prove

$$x \odot (x \to y) = \min(x, y). \tag{5.19}$$

As a matter of fact, by (5.16) the operation $\odot$ is monotonic. Hence, a number $w \in [0, 1]$ coincides with $x \odot t$ for some $t \in [0, 1]$ if and only if $w \in [0, x]$. An application of (5.18) yields the identities $x \odot (x \to y) = x \odot \max\{t \mid x \odot t \le y\} = \max\{x \odot t \mid x \odot t \le y\} = \max\{w \mid w \le y,\ \ w \in [0, x]\} = \min(x, y)$, which completes the proof of (5.19).

Using (5.19), iterated applications of (5.5) and (5.9) yield the following identities: $\max(x, y) {=} \neg \min(\neg x, \neg y) {=} \neg(\neg x \odot (\neg x \to \neg y)) = \neg\neg(\neg x \to \neg(\neg x \to \neg y)) = \neg x \to \neg(\neg x \to \neg y) = (\neg x \to \neg y) \to x = (y \to x) \to x$. Hence

$$(y \to x) \to x = \max(x, y). \tag{5.20}$$

We next prove

$$x \odot y = \neg(\neg x \oplus \neg y) \quad \text{and} \quad x \oplus y = \neg(\neg x \odot \neg y). \tag{5.21}$$

Combining (5.9) and (5.5) we may write $x \oplus y = \neg x \to y = \neg y \to x = \neg(\neg y \odot \neg x)$. The identity $x \odot y = \neg(\neg x \oplus \neg y)$ similarly follows.

To prove

$$x < y \ \Rightarrow\ \neg x > \neg y, \tag{5.22}$$

we argue by contradiction. Assume, on the contrary, $x < y$ and $\neg x \le \neg y$. By (5.2), $\neg x \ge \neg y$, whence $\neg x = \neg y$. By (5.5), $y = \neg\neg y = \neg\neg x = x$, which is impossible.

Recalling that $z \ominus y = z \odot \neg y$ we next prove

$$y < z \quad \Rightarrow \quad (z \ominus y > 0 \quad \text{and} \quad z = y \oplus (z \ominus y)). \tag{5.23}$$

Arguing by contraposition, $z \odot \neg y = 0$ implies $1 = \neg(z \odot \neg y) = z \to y$, whence $z \le y$. In other words, $y < z \Rightarrow z \ominus y > 0$. For the second implication, from $y < z$, an application of (5.9) and (5.20) yields $y \oplus (z \odot \neg y) = \neg y \to \neg(z \to y) = (z \to y) \to y = \max(z, y) = z$. This concludes the proof of (5.23).

Further elementary properties of the $\to$ operation. For any $\epsilon \in [0, 1]$ we have:

$$\text{If} \quad y < 1 \quad \text{and} \quad \epsilon > 0 \quad \text{then} \quad y \oplus \epsilon > y, \tag{5.24}$$

and

$$\text{If} \quad z > 0 \quad \text{and} \quad \epsilon < 1 \quad \text{then} \quad z \odot \epsilon < z. \tag{5.25}$$

We first prove (5.24). By (5.16), the continuous function $\odot$ is monotonically increasing. The range of the map $z \in [y, 1] \mapsto z \ominus y$ is the interval $[0, \neg y]$. As a consequence, for any $0 < \epsilon \le \neg y$ there is $z = z_\epsilon$ satisfying $z \ominus y = \epsilon$ and $z > y$. From (5.23) it follows that $y \oplus \epsilon = y \oplus (z \ominus y) = z > y$. From the commutativity and monotonicity properties of the $\oplus$ operation respectively proved in (5.10) and (5.17), we obtain that the inequality $y \oplus \epsilon > y$ actually holds for all $0 < \epsilon \in [0, 1]$. This settles (5.24).

To prove (5.25), an application of (5.21) yields $z \odot \epsilon = \neg(\neg z \oplus \neg\epsilon)$. Then (5.24) yields $\neg z \oplus \neg\epsilon > \neg z$, since $\neg z < 1$ and $\neg\epsilon > 0$. A final application of (5.22) yields the desired conclusion. □

5.3 The Logic Ł$_\infty$ of Continuous [0, 1]-Valued Events

Definition 5.1 An *MV-algebra* is a structure $A = \{A, 0, \neg, \oplus\}$ with $\neg$ a unary and $\oplus$ a binary operation, satisfying the following equations:

MV1) $x \oplus (y \oplus z) = (x \oplus y) \oplus z$

MV2) $x \oplus y = y \oplus x$

MV3) $x \oplus 0 = x$

MV4) $\neg\neg x = x$

MV5) $x \oplus \neg 0 = \neg 0$

MV6) $\neg(\neg x \oplus y) \oplus y = \neg(\neg y \oplus x) \oplus x$.

Theorem 5.2 *Let $\to$ be a [0, 1]-valued continuous map on $[0, 1]^2$ satisfying conditions (*)-(**) for all $x, y, z \in [0, 1]$. Let the derived operations $\neg, \oplus$ be defined by $\neg x = x \to 0$, $x \oplus y = \neg x \to y$. Then the structure $A_\to = \{[0, 1], 0, \neg, \oplus\}$ is an MV-algebra.*

Proof We use the equations proved in the previous section:

MV1) follows from (5.11).

MV2) is proved in (5.10).

MV3) follows from (5.7) and (5.10): $x \oplus 0 = 0 \oplus x = \neg 0 \to x = 1 \to x = x$.

MV4) is proved in (5.5).

MV5) is proved by writing $x \oplus \neg 0 = \neg 0 \oplus x = \neg\neg 0 \to x = 0 \to x = 1$.

MV6) follows from (5.20). □

MV-algebras stand to Łukasiewicz logic as boolean algebras stand to boolean logic. In the final section on bibliographical notes it is mentioned that the commutativity axiom MV2) is a consequence of the remaining five axioms.

Definition 5.3 The *underlying order* $\leq$ of an MV-algebra $A = (A, 0, \neg, \oplus)$ (or $\leq_A$ if required by the context for definiteness) is defined by $x \leq y \Leftrightarrow \neg x \oplus y = \neg 0 = 1$.

An *ideal* I of A is a subset of A containing 0, *closed under minorants* (i.e., $y \in I$ and $x \leq y \Rightarrow x \in I$), and under applications of the $\oplus$ function (i.e., $x, y \in I \Rightarrow x \oplus y \in I$).

An MV-algebra A is *simple* if its only ideal is the singleton $\{0\}$.

Every MV-algebra A in this paper will satisfy $0 \neq 1$. For short, A is *proper*. Correspondingly, every ideal $\mathfrak{i}$ of A will be *proper*, in the sense that $\mathfrak{i} \neq A$. It is easy to see that ideals of MV-algebras are the same as kernels of homomorphisms.

Definition 5.4 The *standard boolean algebra* is the structure $\{\{0, 1\}, 0, \neg, \oplus\}$, where $\neg x = 1 - x$ and $x \oplus y = \min(1, x + y)$.

In a similar way, the *standard MV-algebra* is the structure $\{[0, 1], 0, \neg, \oplus\}$, where $\neg x = 1 - x$ and $x \oplus y = \min(1, x + y)$.

A routine verification shows that the underlying order of the standard MV-algebra coincides with the natural order of the unit interval [0, 1].

Remark 5.5 Out of respect for Boole, boolean disjunction is understood as the (truncated addition) *monoidal* operation $\oplus$, rather than $\vee$. Boole wrote + and only defined disjunction for incompatible elements. Later it was shown that this partial structure can be naturally extended to what we today understand as a boolean algebra. As proved by Chang, boolean algebras coincide with MV-algebras whose *monoidal* operation $\oplus$ satisfies the idempotence equation $x \oplus x = x$.

Theorem 5.6 *Let* $\to$ *be a* $[0, 1]$*-valued continuous operation on* $[0, 1]^2$ *satisfying conditions (*)-(**) for all* $x, y, z \in [0, 1]$. *Let the MV-algebra* $A_\to = \{[0, 1], 0, \neg, \oplus\}$ *be as defined in Theorem 5.2. Then there is an isomorphism* ϕ *of* $A_\to$ *onto the standard MV-algebra* $[0, 1]$.

Proof *Claim (a):* Let r, s lie in the open interval $[0, 1] \setminus \{0, 1\}$. Then the identity $r = \neg s$ holds if and only if $r \oplus s = 1$ and for all $\epsilon > 0$, $r \oplus (s \ominus \epsilon) < 1$.

($\Rightarrow$) By hypothesis, $r \oplus s = r \oplus \neg r = \neg r \to \neg r = 1$. Furthermore, for all $\epsilon > 0$ with $\epsilon \leq s$, the result proved in (5.25) yields the inequality $\neg r \ominus \epsilon = \neg r \odot \neg \epsilon < \neg r$, since, by hypothesis, $\neg r > 0$ and $\neg \epsilon < 1$. Therefore, $r \oplus (s \ominus \epsilon) = (r \oplus (\neg r \ominus \epsilon)) = \neg r \to (\neg r \ominus \epsilon) < 1$.

($\Leftarrow$) We will tacitly use the monotonicity properties of $\oplus$ established in (5.17). By (5.16) and (5.22), $r \ominus \epsilon = r \odot \neg \epsilon$ is monotonically decreasing for increasing ϵ. The hypothesis $r \oplus s = 1$ means $1 = \neg r \to s$, i.e., $\neg r \leq s$. Conversely, the assumption $r \oplus (s \ominus \epsilon) = \neg r \to (s \ominus \epsilon) < 1$ for all $\epsilon > 0$ implies $\neg r > (s \ominus \epsilon)$ for all $\epsilon > 0$. Since $\ominus$ is continuous and $s \ominus 0 = s$ then $\neg r \geq s$. Having thus proved $\neg r = s$, the desired conclusion now follows from (5.5). This settles Claim (a).

Since $\oplus$ is associative, for any $x \in A_{\to}$ and $k = 1, 2, \ldots$ we may unambiguously use the notation $k.x = \underbrace{x \oplus \cdots \oplus x}_{k \text{ summands}}$ and $0.x = 0$.

Claim (b): Fix $n = 1, 2, \ldots$. Then the equation

$$(n-1).r = \neg r. \tag{5.26}$$

has a unique solution in the underlying universe [0, 1] of the MV-algebra $A_{\to}$.

Trivially, for $n = 1$ the only possible solution is $r = 1$. Next let us arbitrarily fix $n = 2, 3, \ldots$. The range of the monotonically increasing continuous function $y \in [0,1] \mapsto n.y \in [0,1]$ coincides with [0, 1]. As a consequence, the equation (5.26) in the unknown r has at least one solution in the underlying universe [0, 1] of $A_{\to}$. There remains to prove that the solution is unique.

For contradiction, assume that both r_1 and r_2 satisfy (5.26), say with $r_1 < r_2$. By direct inspection of (5.26) neither r_1 nor r_2 belong to $\{0, 1\}$, whence we can write

$$0 < r_1 < r_2 < 1 \quad \text{and} \quad (n-1).r_i = \neg r_i, \quad (i = 1, 2).$$

Turning now to Claim (a), in the implication $r = \neg s \Rightarrow r \oplus (s \ominus \epsilon) < 1$ for all $\epsilon > 0$, let r be replaced by $(n-1).r_i$ and s be replaced by r_i, $(i = 1, 2)$. Then (5.26) yields for $i = 1, 2$,

$$(n-1).r_i = \neg r_i \quad \Rightarrow \quad (n-1).r_i \oplus (r_i \ominus \epsilon) < 1 \Rightarrow (n-1).r_i \; < 1.$$

From the inequalities $(n-1).r_i < 1$ and $r_1 < r_2$, the monotonicity of $\oplus$ proved in (5.17) yields $(n-1).r_1 \leq (n-1) \cdot r_2$. Again by (5.26), this is equivalent to $\neg r_1 \leq \neg r_2$. From (5.22) it follows that $r_1 \geq r_2$, which is impossible. Claim (b) is settled.

In view of Claim (b), for any $n = 1, 2, \ldots$ we may now define the element $\mathsf{a}_{1/n} \in [0, 1]$ by

$$\mathsf{a}_{1/n} = \text{ the unique element } a \in A_{\to} \text{ satisfying } (n-1).a = \neg a \text{ in (5.26)}. \tag{5.27}$$

Recalling conditions (*) and (5.8) we have the equivalences

$$x \leq_{A_{\to}} y \quad \Leftrightarrow \quad x \to y = 1 \quad \Leftrightarrow \quad x \leq y. \tag{5.28}$$

Therefore, the underlying order $\leq_{A_{\to}}$ over the unit interval [0, 1] coincides with the natural order $\leq$. In particular,

$$\leq_{A_{\to}} \text{ is } \textit{complete} \text{ over } [0,1], \tag{5.29}$$

in the sense that every nonempty subset T of $[0, 1]$ has a least upper bound $\sup T$ with respect to $\leq_{A_\rightarrow}$ coinciding with the least upper bound of T with respect to the natural order. Mutatis mutandis, the same holds for $\inf T$. We will henceforth write $\leq$ instead of $\leq_{A_\rightarrow}$ without danger of ambiguity.

Claim (c): The MV-algebra $A_\rightarrow$ is simple.

Arguing by way of contradiction, suppose $A_\rightarrow$ is not simple, i.e., $A_\rightarrow$ has an ideal J different from $\{0\}$. There is $v > 0$ in $A_\rightarrow$ such that either $J = [0, v]$ or $J = [0, v] \setminus \{v\}$. The element

$$j^* = \sup J$$

(existing by (5.29)) satisfies the inequality $j^* > 0$. Since $0 \oplus 0 = 0$ and by (5.27) $\mathsf{a}_{1/2} \oplus \mathsf{a}_{1/2} = 1$, the monotonicity property (5.24) of the continuous function

$$z \in [0, \mathsf{a}_{1/2}] \mapsto z \oplus z \in [0, 1]$$

yields an element $h \leq \mathsf{a}_{1/2}$ such that $h \oplus h = j^*$. Necessarily, $h > 0$ and $h < j^*$, by (5.24). By definition of j^* there is $j \in J$ with $h \leq j$. Thus h belongs to J and so does $j^* = h \oplus h$. Since J is a proper subset of $[0, 1]$ then $j^* < 1$. Another application of (5.24) yields $j^* < j^* \oplus j^* \in J$, against the definition of j^*. We have just proved that the MV-algebra $A_\rightarrow$ is simple, thus settling Claim (c).

To conclude the proof of Theorem 5.6 we argue as follows: Chang's characterization of simple MV-algebras (see TheBook, 3.5.1 and p.76), yields a subalgebra A of the standard MV-algebra $[0, 1]$, along with an isomorphism ϕ of $A_\rightarrow$ onto A. For each $n = 1, 2, \ldots,$ the unique solution $\mathsf{a}_{1/n}$ of (5.27) is sent by the isomorphism ϕ to the unique solution in A of the same equation. Since A is a subalgebra of the standard MV-algebra, then such unique solution coincides with the rational $1/n$. More generally, for each $k = 1, \ldots, n$ the positive integer multiple $\mathsf{a}_{k/n}$ of any element $\mathsf{a}_{1/n}$ is mapped by ϕ to the corresponding multiple k/n of $1/n$. We have thus shown that the range of ϕ contains $\mathbb{Q} \cap [0, 1]$.

There remains to prove that *the range of ϕ coincides with* $[0, 1]$. To this purpose, let us pick an irrational $\xi \in [0, 1]$, with the intent of showing that $\xi = \phi(x)$ for some $x \in A_\rightarrow$. Let $I_1 \supseteq I_2 \supseteq \ldots$ be a descending sequence of closed intervals in $[0, 1]$ with rational vertices such that $\bigcap_m I_m = \{\xi\}$. Since the isomorphism ϕ is order preserving, let $J_1 \supseteq J_2 \supseteq \ldots$ be the corresponding sequence of intervals with vertices of the form $\mathsf{a}_{k/n}$ such that $\phi(J_m) = I_m$. Recalling (5.29), from the completeness of the natural order of the domain $[0, 1]$ of the monotone bijection ϕ it follows that $\bigcap_n J_n$ is nonempty—indeed it is a singleton. We conclude that ξ belongs to the range of ϕ, and ϕ is the desired isomorphism of $A_\rightarrow$ onto the standard MV-algebra $[0, 1]$.

The proof of Theorem 5.6 is now complete. □

Corollary 5.7 *The bijection $\phi : [0, 1] \rightarrow [0, 1]$ of Theorem 5.6 is uniquely determined.*

Proof Combine Theorem 5.6 with TheBook, Corollary 7.2.6. □

As explained in TheBook §4, MV-algebras are the Lindenbaum algebras of infinite-valued Łukasiewicz logic $Ł_\infty$. Łukasiewicz implication $\to_{Ł_\infty}$ is the $[0, 1]$-valued operation on $[0, 1]^2$ defined by

$$x \to_{Ł_\infty} y = \min(1, 1 - x + y). \tag{5.30}$$

Corollary 5.8 (Characterization of Łukasiewicz implication) *For any* $[0, 1]$*-valued map* $\to$ *on* $[0, 1]^2$ *satisfying (*) and (**) there is an order-preserving, (automatically continuous and unique)* bijection ψ *of the unit real interval* $[0, 1]$ *onto itself such that, for all* $x, y \in [0, 1]$, $x \to y = \psi^{-1}(\min(1, 1 - \psi(x) + \psi(y)))$. □

The results of this chapter may be summarized as follows:

$Ł_\infty$ is the only logic arising from a continuous $[0, 1]$-valued function $\to$ having the two properties (*) and (**) of what is usually understood as an implication on a partially ordered set of truth-values with a top and a bottom element.

Historical and bibliographical notes. The first four axioms in Łukasiewicz's original list ([3, p. 144], in Polish notation) for his infinite-valued Łukasiewicz logic $Ł_\infty$ are as follows, for arbitrary $Ł_\infty$-formulas α, β, γ:

$$\alpha \to (\beta \to \alpha)$$
$$(\alpha \to \beta) \to ((\beta \to \gamma) \to (\alpha \to \gamma))$$
$$((\alpha \to \beta) \to \beta) \to ((\beta \to \alpha) \to \alpha)$$
$$(\neg\beta \to \neg\alpha) \to (\alpha \to \beta).$$

Also see TheBook, Sect. 4. The fifth axiom

$$((\alpha \to \beta) \to (\beta \to \alpha)) \to (\beta \to \alpha)$$

occurring in Łukasiewicz's original list was independently proved to be a consequence of the first four by Chang and Meredith in [9]. See TheBook, p.102, for bibliographical details. The derivability of the fifth Łukasiewicz axiom from the remaining four is today a testbed for automated theorem proving. See [2, Prologue]

MV-algebras were introduced by C.C. Chang in [4]. His long list of axioms can be equivalently reduced to the six equations in Definition 5.1. For a proof see the second chapter of TheBook. Furthermore, in [7] it is shown that the commutativity axiom $x \oplus y = y \oplus x$ follows from the remaining five axioms. In [5] Chang proved his completeness theorem, stating that if an equation holds in the standard MV-algebra $[0, 1]$ then it holds in all MV-algebras. With reference to our remark in 5.5, in [4, 1.16-1.17] Chang proved that boolean algebras are the same as *idempotent* MV-algebras, i.e., MV-algebras whose *monoidal* operation $\oplus$ satisfies the idempotence equation

$x \oplus x = x$. See [10] for an elementary (longer) proof of Theorem 5.6, avoiding Chang's characterization of simple MV-algebras, i.e., Theorem 3.5.1 in TheBook.

Corollary 5.8 is known as the Smets-Magrez theorem. Baczyński gives a proof in [1, Theorem 2.4.20] building on the theory of t-norms, [8], and on many intermediate results by several researchers. See [1, p. 65] for details.

Looking over the subject of Łukasiewicz logic as it stands today, it is perhaps appropriate to recall here the following introductory remark by de Finetti in his 1936 paper [6], more than twenty years before the introduction of MV-algebras:

> Il n'y a pas de doute possible, depuis les belles recherches de MM. Łukasiewicz, Reichenbach, Mazurkiewicz, etc., que le calculs des probabilités peut être considéré comme une logique à plusieurs valeurs (précisément: à une échelle continue de valeurs), et que ce point de vue est le plus propre à éclaicir dans les fondements la notion et la logique de la probabilité. Mais ce but est loin d' être atteint par la seule conclusion, de nature purement formelle, que le calcul des probabilités est une logique à plusierurs valeurs: cette conclusion n'est utile que comme point de départ, elle ne constitue pas une façon de résoudre le problème, mais seulement une façon heureuse de le poser nettement. Je crois donc utile d'examiner, comparer et discuter les diverses interprétations qu'on peut donner de cette même théorie des probabilités conçue comme logique à plusieurs valeurs.

References

1. Baczyński, M., & Balasubramaniam, J. (2008). *Fuzzy implications*. Springer.
2. Bonacina, M. P. (2022). Set of support, demodulation, paramodulation: A historical perspective. *Journal of Automated Reasoning, 66*, 463–497.
3. Borkowski, L., & Łukasiewicz, J. (Ed.) (1970). *Selected works. Studies in logic and the foundations of mathematics*. Amsterdam and Polish Scientific Publishers.
4. Chang, C. C. (1958). Algebraic analysis of many valued logics. *Transactions of the American Mathematical Society, 88*, 467–490.
5. Chang, C. C. (1959). A new proof of the completeness of the Łukasiewicz axioms. *Transactions of the American Mathematical Society, 93*, 74–90.
6. de Finetti, B. (1935, 1936). La logique de la probabilité. In *Actes du congrés international de philosophie scientifique, Sorbonne, Paris* (pp. 565–573). Hermann et C.ie Éditeurs.
7. Kolařík, M. (2013). Independence of the axiomatic system for MV-algebras. *Mathematica Slovaca, 63*, 1–4.
8. Klement, E. P., Mesiar, R., & Pap, E. (2000). *Triangular norms, Trends in Logic* (Vol. 8). Kluwer.
9. Meredith, C. A. (1958). The dependence of an axiom of Łukasiewicz. *Transactions of the American Mathematical Society, 87*, 54.
10. Mundici, D. (2020). What the Łukasiewicz axioms mean. *The Journal of Symbolic Logic, 85*(3), 906–917.

Chapter 6
Probabilistic Consistency of [0, 1]-Events

In this chapter, Theorem 3.8 will be extended from boolean $\{0, 1\}$-events to continuous $[0, 1]$-valued events. Let $\mathsf{hom}(A)$ denote the set of homomorphisms of A into the standard MV-algebra $[0, 1]$. For any MV-algebra A and set $E = \{h_1, \ldots, h_m\} \subseteq A$, a map $\beta \colon E \to [0, 1]$ is said to be (probabilistically) *inconsistent* if

$$\text{there is } s \colon E \to \mathbb{R} \text{ such that for any } \eta \in \mathsf{hom}(A), \sum_{j=1}^{m} s(h_j)(\beta(h_j) - \eta(h_j)) < 0.$$

This definition extends to MV-algebras the game-theoretic reformulation of boolean probabilistic consistency: Intuitively, Betty can devise a "bet" $(s_1, \ldots, s_m) \in \mathbb{R}^m$ on the continuous events $(h_1, \ldots, h_m)$ in E that "bankrupts" Bookie in any "possible world" η. As in the boolean case, in Theorem 6.3 we will show that β is consistent if and only if it is extendable to a *state* of A, i.e., to a normalized function $\sigma \colon A \to [0, 1]$ having the additivity property. The main properties of the space of states of MV-algebras will be derived from their counterparts for *unital ℓ-groups*, i.e., lattice-ordered abelian groups with a distinguished (strong order) unit. The link between these two classes of structures is provided by an equivalence between the categories of unital ℓ-groups and MV-algebras, known as the Γ functor.

6.1 States and the MV-Algebraic Consistency Theorem

Definition 6.1 A *state* of an MV-algebra A is a function $\sigma \colon A \to [0, 1]$ such that $\sigma(1) = 1$ and $\sigma(a \oplus b) = \sigma(a) + \sigma(b)$ whenever $a \odot b = 0$. $\mathsf{S}(A)$ denotes the state space of the MV-algebra A equipped with the restriction topology of the Tychonov cube $[0, 1]^A$.

We also let $\mathsf{hom}(A)$ denote the set of homomorphisms of A into the standard MV-algebra $[0, 1]$.

D. Mundici, *Algebraic Probabilistic Consistency*, Trends in Logic 69,
https://doi.org/10.1007/978-3-031-98337-5_6

Observe that when A is a boolean algebra the range of any MV-algebraic homomorphism $\eta\colon A \to [0,1]$ necessarily coincides with $\{0,1\}$, and the $\odot$ operation coincides with $\wedge$. Hence the present definitions of states and homomorphisms of A agree with the boolean algebraic definitions of $\mathsf{S}(A)$ and $\mathsf{hom}(A)$ given in the first part of the book.

Definition 6.2 Let A be an MV-algebra and $E = \{h_1, \dots, h_m\}$ a subset of A. Then a function $\beta\colon E \to [0,1]$, is said to be (probabilistically) *consistent* if

$$\text{for any } s\colon E \to \mathbb{R} \text{ there is } \eta \in \mathsf{hom}(A) \text{ with } \sum_{j=1}^{m} s(h_j)\cdot(\beta(h_j) - \eta(h_j)) \geq 0. \tag{6.1}$$

When A is a boolean algebra Definition 6.2 agrees with Definition 3.7.

Theorem 6.3 (MV-algebraic probabilistic theorem) *For any MV-algebra A and subset $E = \{h_1, \dots, h_m\}$ of A, a function $\beta\colon E \to [0,1]$ is consistent if and only if it can be extended to a state of A.*

The proof is deferred until Sects. 6.4–6.5. The necessary material will be provided in the next few sections. When A is a boolean algebra, Theorems 6.3 and 3.8 coincide.

Note that the "logical" consistency of β in Łukasiewicz logic amounts to the extendability of β to a homomorphism of A into $[0,1]$. To avoid confusion, unless otherwise specified, "consistency" will always be understood in the probabilistic sense.

First properties of the states of an MV-algebra. For any set X, the Tychonov cube $[0,1]^X$ is the set of functions $f\colon X \to [0,1]$, endowed with the product topology, making it into a compact Hausdorff space. For any subset Y of $[0,1]^X$, we write

$$\mathsf{conv}\, Y \quad\text{and}\quad \mathsf{cl}\, Y$$

respectively for the *convex hull* and for the *closure* of Y in $[0,1]^X$. For K a convex subset of $[0,1]^X$,

$$\mathsf{ext}\, K \text{ is shorthand for the set of all } \textit{extreme points} \text{ of } K.$$

By Corollary 1.2.14 in TheBook, any MV-algebra A has a maximal ideal $\mathfrak{m}$. By Proposition 1.2.10 in TheBook, $A/\mathfrak{m}$ is *simple*, in the sense that the only ideal of the quotient $A/\mathfrak{m}$ is the zero ideal $\{0\}$. By Theorem 3.5.1 in TheBook, there is an embedding η of $A/\mathfrak{m}$ into $[0,1]$. The composition of η with the quotient map $a \in A \mapsto a/\mathfrak{m} \in A/\mathfrak{m}$ yields an element of $\mathsf{hom}(A)$. We conclude that $\mathsf{hom}(A)$ is nonempty, whence so is $\mathsf{S}(A)$.

A state ρ of A is *extreme* if $\rho \notin \mathsf{conv}(\mathsf{S}(A) \setminus \{\rho\})$. In what follows

$$\mathsf{ext}\, \mathsf{S}(A) \text{ denotes the set of extreme states of } A.$$

For later use, in the following proposition we list some elementary consequences of the definition:

Proposition 6.4 *In any MV-algebra A we have:*

(i) $\mathsf{S}(A)$ *is a convex subset of the vector space* $\mathbb{R}^A$.

(ii) Any $\eta \in \mathsf{hom}(A)$ *is a state of A, and so is any convex combination of states.*

(iii) Equipped with the restriction topology of $[0, 1]^A$, *both* $\mathsf{S}(A)$ *and* $\mathsf{hom}(A)$ *are (closed and) compact Hausdorff subspaces of* $[0, 1]^A$. □

At the end of Chap. 1 it was noted that states of MV-algebras are the counterpart of states in lattice-ordered abelian groups with a distinguished strong order unit, as defined in Definition 6.5. Indeed, many preliminary results on states in MV-algebras can be proved as direct consequences of their counterparts for unital ℓ-groups. Since all these results are proved in Goodearl's comprehensive textbook "Partially Ordered Abelian Groups with Interpolation", [1] there is no need to duplicate their proofs here. As a matter of fact, using the categorical equivalence Γ between unital ℓ-groups and MV-algebras (see Definition 6.7) all these results directly translate into the language of MV-algebras and their states. The reader is only assumed to have some familiarity with the main properties of the Γ functor, as detailed in Chaps. 2 and 7 of TheBook.

Definition 6.5 By a *partially ordered abelian group* we mean an abelian group G, together with a distinguished subset $G^+ \subseteq G$ (the *positive cone* of G) such that

$$G^+ + G^+ = G^+, \quad G^+ - G^+ = G, \quad \text{and} \quad G^+ \cap -G^+ = \{0\}.$$

Upon setting $x \leq y$ iff $y - x \in G^+$ we equip G with a partial order. When this partial order is a lattice, G is said to be a *lattice-ordered abelian group*, ℓ-group for short.

A *unital* ℓ-group is a pair (G, u) where G is an ℓ-group and $u \in G$, the (distinguished strong order-) unit of G has the following property: For each $x \in G$ there is $n = 1, 2, \ldots$ such that $x \leq nu$. The standard unital ℓ-group is $(\mathbb{R}, 1)$ where $\mathbb{R}$ is the additive group of real numbers with the natural order.

A *state* of a unital ℓ-group (G, u) is a group homomorphism τ of (G, u) into $(\mathbb{R}, 1)$ such that $\tau(G^+) \subseteq \mathbb{R}^+$ and $\tau(u) = 1$. $\mathsf{S}(G, u)$ denotes the convex set of states of (G, u). A state $\tau \in \mathsf{S}(G, u)$ is *extreme* if $\tau \notin \mathsf{conv}(\mathsf{S}(G, u) \setminus \{\tau\})$. We denote by $\mathsf{ext}\,\mathsf{S}(G, u)$ the set of extreme states of (G, u).

A *unital* ℓ*-homomorphism* $\phi\colon (G, u) \to (G', u')$ is an order preserving group homomorphism of G into G' that maps u into u'. We let $\mathsf{hom}(G, u)$ denote the set of unital ℓ-homomorphisms of (G, u) into $(\mathbb{R}, 1)$. Evidently, $\mathsf{hom}(G, u) \subseteq \mathsf{S}(G, u)$.

A first ingredient of the proof of the consistency theorem for MV-algebras is the following proposition, dealing with the mutual relationships between $\mathsf{S}(G, u)$ and $\mathsf{hom}(G, u)$:

[1] Cited as [8] in the bibliography at the end of this chapter.

Theorem 6.6 *For any unital* ℓ*-group* (G, u) *we have:*

(i) $\mathsf{S}(G, u)$ *is a compact Hausdorff subspace of* $\mathbb{R}^G$*, and* $\mathsf{hom}(G, u)$ *is a compact Hausdorff subspace of* $\mathsf{S}(G, u)$.

(ii) $\mathsf{hom}(G, u) = \mathsf{ext}\, \mathsf{S}(G, u)$.

(iii) $\mathsf{S}(G, u)$ *is the closure in the Tychonov cube* $[0, 1]^G$ *of the set* $\mathsf{conv}\, \mathsf{hom}(G, u)$ *of convex combinations of homomorphisms of* (G, u) *into* $(\mathbb{R}, 1)$*. In symbols,*

$$\mathsf{S}(G, u) = \mathsf{cl}\, \mathsf{conv}\, \mathsf{hom}(G, u). \tag{6.2}$$

(iv) Let (G, u) *and* (H, u) *be unital* ℓ*-groups with the same (distinguished) unit. If* G *is an* ℓ*-subgroup of* H *then every state of* (G, u) *extends to a state of* (H, u).

Proof (i) The compactness of $\mathsf{S}(G, u)$ is proved in [8, 6.2, p. 96]. The compactness of the closed set $\mathsf{hom}(G, u) \subseteq \mathsf{S}(G, u)$ then immediately follows.

(ii) For a proof see [8, 12.18, p. 203].

(iii) Combine (i)-(ii) with the Krein-Milman theorem, [8, 5.17, p. 85].

(iv) This is a particular case of [8, Corollary 4.3, p. 62]. □

6.2 The Γ Functor from Unital ℓ-Groups to MV-Algebras

The above results on unital ℓ-groups directly transfer to MV-algebras due to the preservation properties of the following functor:

Definition 6.7 Let (G, u) be a unital ℓ-group. On the interval $[0, u] \subseteq G$ let the operations $\neg$ and $\oplus$ be defined by $\neg x = u - x$ and $x \oplus y = (x + y) \wedge u$. Let $\Gamma(G, u) = ([0, u], 0, \neg, \oplus)$. Furthermore, for every unital ℓ-group homomorphism $\eta \colon (G, u) \to (H, v)$ let $\Gamma(\eta)$ be the restriction of η to $[0, u]$.

Theorem 6.8 (Chaps. 2 and 7 in $\mathsf{TheBook}$) Γ *is a categorical equivalence between the category of unital* ℓ*-groups with unital* ℓ*-group homomorphisms, and the category of MV-algebras with their homomorphisms.* □

In every MV-algebra $A = (A, 0, \neg, \oplus)$ we define

$$1 = \neg 0, \quad a \odot b = \neg(\neg a \oplus \neg b), \quad a \to b = \neg a \oplus b, \quad \text{and} \quad x \ominus y = x \odot \neg y. \tag{6.3}$$

In the interpretation of MV-terms, $\neg$ is more binding than $\odot$, and the latter is more binding than $\oplus$.

Proposition 6.9 *For every state σ of an MV algebra A and $a, b \in A$ we have:*

(i) *If $a \leq b$ then $\sigma(a) \leq \sigma(b)$.*

(ii) $\sigma(a) + \sigma(b) = \sigma(a \oplus b) + \sigma(a \odot b)$ *for all $a, b \in A$.*

(iii) $\sigma(0) = 0$.

(iv) $\sigma(\neg a) = 1 - \sigma(a)$.

Proof (i) Upon setting $c = \neg a \odot b$ we have $a \odot c = 0$, and $a \oplus c = a \oplus (\neg a \odot b) = \neg(a \oplus \neg b) \oplus a = \neg(b \oplus \neg a) \oplus b = \neg 1 \oplus b = b$. As a consequence, $\sigma(b) = \sigma(a \oplus c) = \sigma(a) + \sigma(c) \geq \sigma(a)$.

(ii) Trivially,

$$((a \oplus b) \odot \neg b) \odot (a \odot b) = 0.$$

Recalling the definition of $\ominus$ in (6.3), a proof of the equation

$$(a \odot b) \oplus ((a \oplus b) \odot \neg b) = a$$

can be obtained from TheBook, (1.15), p. 27, replacing a by x and b by $\neg y$. From these two equations we obtain $\sigma(a) = \sigma((a \oplus b) \odot \neg b) + \sigma(a \odot b)$ and, symmetrically, $\sigma(b) = \sigma((a \oplus b) \odot \neg a) + \sigma(a \odot b)$. Next we observe that the standard MV-algebra $[0, 1]$ satisfies the following equations:

$$((a \oplus b) \odot \neg b) \odot ((a \oplus b) \odot \neg a) = 0 = ((a \oplus b) \odot \neg b \;\oplus\; (a \oplus b) \odot \neg a) \odot a \odot b$$

$$\text{and} \quad (a \oplus b) \odot \neg b \;\oplus\; (a \oplus b) \odot \neg a \;\oplus\; a \odot b = a \oplus b.$$

To prove that these equations hold in $[0, 1]$ one argues by cases, according as $a \leq b$ or $a > b$. By Chang's completeness theorem (see TheBook, 2.5.3), all these equations are also valid in A. As a consequence we obtain:

$$\begin{aligned}
\sigma(a) + \sigma(b) &= \sigma((a \oplus b) \odot \neg b) + \sigma((a \oplus b) \odot \neg a) + \sigma(a \odot b) + \sigma(a \odot b) \\
&= \sigma((a \oplus b) \odot \neg b \oplus (a \oplus b) \odot \neg a) + \sigma(a \odot b) + \sigma(a \odot b) \\
&= \sigma((a \oplus b) \odot \neg b \oplus (a \oplus b) \odot \neg a \oplus a \odot b) + \sigma(a \odot b) \\
&= \sigma(a \oplus b) + \sigma(a \odot b).
\end{aligned}$$

(iii) The additivity property of σ yields $\sigma(0) = \sigma(0 \oplus 0) = \sigma(0) + \sigma(0)$, since $0 \odot 0 = 0$.

(iv) Again by additivity, from $a \odot \neg a = 0$ we get $\sigma(\neg a) + \sigma(a) = \sigma(a \oplus \neg a) = \sigma(1) = 1$. □

6.3 Further Properties of the States of an MV-Algebra

For any vector spaces V_1 and V_2 and convex subsets $K_1 \subseteq V_1$ and $K_2 \subseteq V_2$, by an *affine isomorphism* between K_1 and K_2 we mean a bijection ϕ of K_1 onto K_2 that preserves convex combinations. If, in addition, K_1 and K_2 are topological vector spaces and ϕ is a homeomorphism, we say that ϕ is an *affine homeomorphism.*

Theorem 6.10 *Let (G, u) be a unital ℓ-group and $A = \Gamma(G, u)$ the MV-algebra corresponding to (G, u), where the equivalence Γ identifies the unit element 1 of A and the (distinguished, strong, order) unit u of (G, u). For any state ρ of (G, u) we then have:*

(i) The restriction $\rho \upharpoonright A$ to the unit interval $[0, u]$ of (G, u) is a state of A.

(ii) Every state of A arises as the restriction to $[0, u]$ of a state of (G, u).

(iii) The restriction map $\rho \mapsto \rho \upharpoonright A$ is an affine homeomorphism of $\mathsf{S}(G, u) \subseteq \mathbb{R}^G$ onto $\mathsf{S}(A) \subseteq [0, 1]^A$, mapping $\mathsf{ext}\,\mathsf{S}(G, u)$ homeomorphically onto $\mathsf{ext}\,\mathsf{S}(A)$.

Proof (i) Let $\rho^{\upharpoonright}$ be shorthand for $\rho \upharpoonright A$. Trivially, $\rho^{\upharpoonright}(u) = u = 1$. To prove that $\rho^{\upharpoonright}$ has the additivity property, by TheBook, 2.1.3(i) we have

$$x + y = (x \oplus y) + (x \odot y) \text{ for all } x, y \in A.$$

Therefore, $\rho^{\upharpoonright}(x) + \rho^{\upharpoonright}(y) = \rho(x) + \rho(y) = \rho(x + y) = \rho(x \oplus y + x \odot y) = \rho(x \oplus y) + \rho(x \odot y) = \rho^{\upharpoonright}(x \oplus y) + \rho^{\upharpoonright}(x \odot y)$, which shows that $\rho^{\upharpoonright}$ is a state of A.

(ii) Following Proposition 2.4.2(iii) in TheBook, let us identify the positive cone G^+ of G with the abelian lattice-ordered monoid M_A of good sequences of A. As the reader will recall from TheBook §§2.2-2.3, for any $p \in G^+$ the *good sequence* $\mathsf{g} = (\mathsf{g}_1, \ldots, \mathsf{g}_n) \in A^n$ of p is defined by stipulating that for each $i = 1, \ldots, n - 1$,

$$\mathsf{g}_i \oplus \mathsf{g}_{i+1} = \mathsf{g}_i \quad \text{and} \quad p = \mathsf{g}_1 + \cdots + \mathsf{g}_n.$$

By TheBook §2.2, the good sequence g of p is uniquely determined up to a finite final tail of zeros. For any $\tau \in \mathsf{S}(A)$ let the function $\bar{\tau} \colon M_A \to \mathbb{R}^+$ be defined by

$$\bar{\tau}(p) = \bar{\tau}(\mathsf{g}) = \tau(\mathsf{g}_1) + \cdots + \tau(\mathsf{g}_n).$$

Claim: The map $\bar{\tau}$ is additive.

It suffices to show $\bar{\tau}(p + a) = \bar{\tau}(p) + \bar{\tau}(a)$ for all $p \in G^+$ and $a \in A$. To this purpose, following TheBook 2.2.4, we define the $(n + 1)$-tuple $\mathsf{g}' = (\mathsf{g}'_1, \ldots, \mathsf{g}'_{n+1}) \in A^{n+1}$ by

$$\begin{aligned} \mathsf{g}'_1 &= \mathsf{g}_1 \oplus a, \\ \mathsf{g}'_2 &= \mathsf{g}_2 \oplus (\mathsf{g}_1 \odot a), \end{aligned}$$

$$\begin{aligned}
\mathsf{g}'_3 &= \mathsf{g}_3 \oplus (\mathsf{g}_2 \odot a),\\
\dots & \quad \dots \quad \dots\\
\mathsf{g}'_n &= \mathsf{g}_n \oplus (\mathsf{g}_{n-1} \odot a),\\
\mathsf{g}'_{n+1} &= \mathsf{g}_n \odot a.
\end{aligned}$$

Arguing as in TheBook §2.3, and recalling our identification $G^+ = M_A$, it follows that g' is the good sequence of $p + a \in G^+$. By Proposition 6.9(ii),

$$\begin{aligned}
\bar{\tau}(a+p) &= \tau(\mathsf{g}'_1) + \tau(\mathsf{g}'_2) + \tau(\mathsf{g}'_3) + \cdots + \tau(\mathsf{g}'_n) + \tau(\mathsf{g}'_{n+1})\\
&= \tau(a) + \tau(\mathsf{g}_1) - \tau(\mathsf{g}_1 \odot a) + \tau(\mathsf{g}_2) + \tau(\mathsf{g}_1 \odot a) - \tau(\mathsf{g}_2 \odot a) + \tau(\mathsf{g}_3)\\
&+ \tau(\mathsf{g}_2 \odot a) - \tau(\mathsf{g}_3 \odot a) + \cdots + \tau(\mathsf{g}_n) + \tau(\mathsf{g}_{n-1} \odot a) - \tau(\mathsf{g}_n \odot a)\\
&+ \tau(\mathsf{g}_n \odot a) = \tau(a) + \tau(\mathsf{g}_1) + \cdots + \tau(\mathsf{g}_n) = \bar{\tau}(a) + \bar{\tau}(\mathsf{g}) = \bar{\tau}(a) + \bar{\tau}(p).
\end{aligned}$$

Our claim is settled.

By Proposition 6.9(iii), $\bar{\tau}(0) = 0$. Now, $G = G^+ - G^+$ and $\bar{\tau}$ is a monoid homomorphism of G^+ into $\mathbb{R}^+$. A state $\bar{\bar{\tau}}$ extending $\bar{\tau} \supseteq \tau$ to $(G, 1) = (G, u)$ is now obtained by setting $\bar{\bar{\tau}}(x - y) = \bar{\tau}(x) - \bar{\tau}(y)$ for all $x, y \in G^+$.

(iii) A direct inspection of the compact topologies of $\mathsf{S}(A)$ and $\mathsf{S}(G, u)$ shows that the maps $\rho \mapsto \rho^{\restriction} = \rho \restriction A$ and $\tau \mapsto \bar{\bar{\tau}}$ are mutually inverse affine homeomorphisms between these two compact Hausdorff spaces. The desired homeomorphism of the compact set of extreme states of $(G, 1) = (G, u)$ onto the compact set of extreme states of A is now obtained by restriction. □

Corollary 6.11 *For any MV-algebra A,* $\mathsf{hom}(A) = \mathsf{ext}\,\mathsf{S}(A)$.

Proof From Theorems 6.6(ii) and 6.10(iii). □

Theorem 6.12 *For every MV-algebra A, the set of states of A coincides with the closure in $[0, 1]^A$ of the set of convex combinations of homomorphisms of A into the standard MV-algebra. In symbols,*

$$\mathsf{S}(A) = \mathsf{cl}\,\mathsf{conv}\,\mathsf{hom}(A). \tag{6.4}$$

Proof Mimicking the proof of Theorem 6.6(iii), and using the affine homeomorphism of Theorem 6.10 and the Krein-Milman theorem, again from [8, 5.17, p.85] we can write:

$$\begin{aligned}
\mathsf{S}(A) &= \mathsf{cl}\,\mathsf{conv}\,\mathsf{ext}\,\mathsf{S}(A)\\
&= \mathsf{cl}\,\mathsf{conv}\,\mathsf{hom}(A), \quad \text{by Corollary 6.11.}
\end{aligned}$$

This completes the proof. □

Proposition 6.13 *Let A and B be MV-algebras with A a subalgebra of B. Then every state ρ of A extends to a state of B.*

Proof Let (G, u) and (H, v) be unital ℓ-groups, respectively corresponding to A and B via the Γ functor. Since Γ is an equivalence, (G, u) is an ℓ-subgroup of (H, v), and $v = u$. By Theorem 6.10(ii)–(iii), ρ uniquely determines a state $\bar{\bar{\rho}}$ of (G, u). By Theorem 6.6(iv), $\bar{\bar{\rho}}$ extends to a state ρ_H of (H, v). A straightforward argument shows that the restriction $\rho_H \restriction A$ is a state of A extending ρ. □

6.4 Proof of the MV-Algebraic Consistency Theorem: First Part

(If β is consistent then β is extendable to a state $\tau \in \mathsf{S}(A)$.) We argue by contraposition. Assuming that no state of A extends β we must find a vector $s = (s_1, \ldots, s_m) \in \mathbb{R}^m$ satisfying the following condition, where s_j is shorthand for $s(h_j)$:

$$\text{for all } \eta \in \mathsf{hom}(A), \quad \sum_{j=1}^{m} s_j(\beta(h_j) - \eta(h_j)) < 0. \tag{6.5}$$

For any state τ of A let the vector $\tau_E \in [0, 1]^m$ be defined by

$$\tau_E = (\tau(h_1), \ldots, \tau(h_m)).$$

Likewise, let the vector $\beta_E \in [0, 1]^m$ be defined by $\beta_E = (\beta(h_1), \ldots, \beta(h_m))$. With $|x|$ denoting the euclidean norm of $x \in \mathbb{R}^m$, our standing hypothesis that no state of A extends β may be rewritten as

$$\text{for all } \ \tau \in \mathsf{S}(A), \qquad |\beta_E - \tau_E| > 0.$$

By Proposition 6.4(iii), the state space $\mathsf{S}(A)$ is a compact subset of the Tychonov cube $[0, 1]^A$. Since the map $\tau \mapsto |\beta_E - \tau_E|$ is continuous there is $\delta > 0$ such that $|\beta_E - \tau_E| > \delta$ for all $\tau \in \mathsf{S}(A)$. More precisely,

$$\text{for some } \delta^* > 0, \ \ \min\{\text{distance}(\beta_E, \tau_E) \mid \tau \in \mathsf{S}(A)\} = \delta^*. \tag{6.6}$$

Let the set $S_E \subseteq [0, 1]^m$ and its translate $\beta_E - S_E$ in $\mathbb{R}^m$ be defined by

$$S_E = \{\tau_E \mid \tau \in \mathsf{S}(A)\}, \qquad \beta_E - S_E = \{\beta_E - \tau_E \ \mid \ \tau \in \mathsf{S}(A)\}.$$

By Proposition 6.4, S_E inherits the properties of convexity and compactness of $\mathsf{S}(A)$. Since by (6.6) the origin does not lie in the closed convex set $\beta_E - S_E$, let $0 \neq x_* \in \mathbb{R}^m$ be the shortest vector in $\beta_E - S_E$ as given by Lemma 2.2(i). Let x_*^T denote the transpose of x_*. By Lemma 2.2(iii)–(v), the hyperplane

$$H = \{z \in \mathbb{R}^m \mid x_*^T z = |x_*|^2/2\}$$

has the following (*strong separation*) property:

> The distance of H from the origin is > 0, the distance from H of any vector $y \in \beta_E - S_E$ is > 0, and $\beta_E - S_E$ and the origin are contained in opposite closed halfspaces with boundary H.

Letting the vector $s \in \mathbb{R}^m$ be given by $s = -x_*$ we then immediately infer that

$$\text{the scalar product } s^T(\beta_E - x) \text{ has a value } < 0 \text{ for all } x \in S_E. \tag{6.7}$$

Intuitively, the vector $s = (s_1, \ldots, s_m)$ *forms an obtuse angle* with every vector in the compact convex set $\beta_E - S_E$. Next let the set $\mathsf{hom}(A)_E \in [0,1]^m$ be defined by

$$\mathsf{hom}(A)_E = \{(\eta(h_1), \ldots, \eta(h_m)) \mid \eta \in \mathsf{hom}(A)\}.$$

Since $\mathsf{hom}(A) \subseteq \mathsf{S}(A)$ then $\mathsf{hom}(A)_E \subseteq S_E$. As a particular case of (6.7),

$$s^T(\beta_E - x) < 0 \quad \text{for all } x \in \mathsf{hom}(A)_E,$$

which amounts to the inconsistency property (6.5) of β. The proof of Theorem 6.3(i)⇒(ii) is complete. □

6.5 Proof of the MV-Algebraic Consistency Theorem: Conclusion

(If β is extendable to some $\sigma \in \mathsf{S}(A)$ then β is consistent.) By way of contradiction assume there is $\sigma \in \mathsf{S}(A)$ extending β and there is $s = (s_1, \ldots, s_m) \in \mathbb{R}^m$ such that

$$\sum_{j=1}^{m} s_j(\beta(h_j) - \eta(h_j)) < 0 \quad \text{for all } \eta \in \mathsf{hom}(A). \tag{6.8}$$

Without loss of generality,

$$s \in [-1, 1]^m. \tag{6.9}$$

By Proposition 6.4(iii), the map

$$\eta \in \mathsf{hom}(A) \mapsto \sum_{j=1}^{m} s_j(\beta(h_j) - \eta(h_j))$$

attains a maximum value ν. By our absurdum hypothesis (6.8) we may write

$$\sum_{j=1}^{m} s_j(\beta(h_j) - \eta(h_j)) \leq \nu < 0 \text{ for all } \eta \in \mathsf{hom}(A). \tag{6.10}$$

Let $\lambda_{j1}, \ldots, \lambda_{jq}$ be arbitrarily chosen real numbers ≥ 0 summing up to 1, and $\eta_{j1}, \ldots, \eta_{jq} \in \mathsf{hom}(A)$. Then an easy verification yields

$$\sum_{j=1}^{m} s_j \left(\beta(h_j) - \sum_{k=1}^{q} \lambda_{jk}\eta_{jk}(h_j) \right) \leq \nu < 0. \tag{6.11}$$

Since $\sum_{k=1}^{q} \lambda_{jk}\eta_{jk}$ is a state of A, for all $\gamma \in \mathsf{conv}\,\mathsf{hom}(A)$ we can write

$$\sum_{j=1}^{m} s_j(\beta(h_j) - \gamma(h_j)) \leq \nu < 0. \tag{6.12}$$

Since by Theorem 6.12, $\mathsf{S}(A) = \mathsf{cl}\,\mathsf{conv}\,\mathsf{hom}(A)$, the set

$$S_E = \{(\tau(h_1), \ldots, \tau(h_m)) \mid \tau \in \mathsf{S}(A)\} \subseteq [0, 1]^m$$

inherits the convexity and compactness properties of $\mathsf{S}(A)$. As a consequence, for all $\tau \in \mathsf{S}(A)$ and $0 < \theta \in \mathbb{R}$ there is a state $\gamma = \gamma_{\tau,\theta} \in \mathsf{conv}\,\mathsf{hom}(A)$ such that

$$|(\tau(h_1), \ldots, \tau(h_m)) - (\gamma_{\tau,\theta}(h_1), \ldots, \gamma_{\tau,\theta}(h_m))| < \theta. \tag{6.13}$$

Thus, in particular, for the state $\sigma \supseteq \beta$ existing by our standing hypothesis, the following holds: For any $\theta > 0$ there is $\gamma = \gamma_\theta \in \mathsf{conv}\,\mathsf{hom}(A)$ such that

$$|(\sigma(h_1), \ldots, \sigma(h_m)) - (\gamma_\theta(h_1), \ldots, \gamma_\theta(h_m))| < \theta. \tag{6.14}$$

Therefore,

$$\begin{aligned}
0 &= \sum_{j=1}^{m} s_j(\beta(h_j) - \sigma(h_j)), \text{ since } \sigma \supseteq \beta \text{ by hypothesis} \\
&= \sum_{j=1}^{m} s_j(\beta(h_j) - \gamma_\theta(h_j) + \gamma_\theta(h_j) - \sigma(h_j)) \\
&= \sum_{j=1}^{m} s_j(\beta(h_j) - \gamma_\theta(h_j)) + \sum_{j=1}^{m} s_j(\gamma_\theta(h_j) - \sigma(h_j)) \\
&\leq \nu + \sum_{j=1}^{m} s_j(\gamma_\theta(h_j) - \sigma(h_j)), \text{ by (6.12)} \\
&\leq \nu + \sum_{j=1}^{m} |s_j| \cdot |\gamma_\theta(h_j) - \sigma(h_j)|
\end{aligned}$$

$$\leq \nu + \sum_{j=1}^{m} 1 \cdot |\gamma_\theta(h_j) - \sigma(h_j)|, \text{ by (6.9)}$$
$$\leq \nu + m\theta, \text{ by (6.14)}.$$

Since $\theta > 0$ is arbitrary, letting now $\theta < -\nu/(2m)$ and recalling (6.10) we obtain

$$0 = \sum_{j=1}^{m} s_j(\beta(h_j) - \sigma(h_j)) < \nu/2 < 0,$$

a contradiction. Having thus completed the proof of the (ii)⇒(i) direction, the proof of Theorem 6.3 is complete. □

The following result generalizes Corollary 3.10 to MV-algebras:

Corollary 6.14 *A* $[0, 1]$*-valued function* f *defined on an MV-algebra* A *is a state if and only if it is* finitely consistent, *in the sense that every finite restriction of* f *is consistent.*

Proof (Compare with the proof of Corollary 3.10.) Let $f : A \to [0, 1]$ be a fixed but otherwise arbitrary finitely consistent function, with the intent of extending f to a state of A. Since every finite restriction β of f is consistent, Theorem 6.3 shows that β is extendable to a state τ of A. The set $S_\beta = \{\tau \in \mathsf{S}(A) \mid \tau \supseteq \beta\}$ of such extensions is a closed nonempty subset of $\mathsf{S}(A)$. Let now the family Σ_f of closed subsets of $[0, 1]^A$ be defined by

$$\Sigma_f = \{S_\beta \mid \beta \text{a finite restriction of} f\}.$$

By direct inspection, Σ_f has the Finite Intersection Property: For any finite set $\{\beta_1, \ldots, \beta_u\}$ of finite restrictions of f, the closed set $\Sigma_{\beta_1} \cap \cdots \cap \Sigma_{\beta_u}$ is nonempty. From the compactness of $\mathsf{S}(A) \subseteq [0, 1]^A$ it follows that the intersection $\bigcap \Sigma_f$ is a singleton. Indeed the only element of $\bigcap \Sigma_f$ is a state coinciding with f.

Conversely, let σ be a state of A. Every finite restriction β of σ is extendable to the state σ. By the converse direction of Theorem 6.3, β is consistent. Thus σ is finitely consistent. □

The proof of the above corollary yields the following more general result, which is also a generalization of Corollary 3.11:

Corollary 6.15 *A* $[0, 1]$*-valued map* f *defined on a (possibly infinite) subset of an MV-algebra* A *is extendable to a state of* A *iff so is every finite restriction of* f *iff every finite restriction of* f *is consistent.* □

Corollary 6.16 (Compactness of Łukasiewicz logic $Ł_\infty$) *A* $[0, 1]$*-valued map* g *defined on a (possibly infinite) subset of an MV-algebra* A *is extendable to a homomorphism of* A *into the standard MV-algebra* $[0, 1]$ *iff so is every finite restriction of* g *iff every finite restriction of* g *is (logically) consistent.* □

6.6 Boole's "Object of Probability" Extended to MV-Algebras

Recall from Sect. 1.6 of Chap. 1 Boole's quotation in [2] concerning "the object of the theory of probabilities": Given the probabilities of any events, of whatever kind, to find the probability of some other event connected with them. For boolean events, in Theorem 1.12 de Finetti proved that the set of possible probabilities of the new events is a closed interval in [0, 1].

To prove the MV-algebraic generalization of this theorem, let A be a fixed but otherwise arbitrary MV-algebra and $\mathsf{S}(A)$ the compact convex set of states of A. For any subset $E = \{h_1, \ldots, h_m\}$ of A, function $\beta\colon E \to [0, 1]$, and element $h \in A$, we will use the notation

$$S_\beta = \{\sigma \in \mathsf{S}(A) \mid \sigma \supseteq \beta\} \quad \text{and} \quad S_\beta(h) = \{\sigma(h) \mid \sigma \in S_\beta\}.$$

In Boole's original language, $S_\beta(h)$ is the set of possible probabilities of the event h given the probabilities $\beta(h_1), \ldots, \beta(h_m)$ of the events $h_1, \ldots, h_m$.

Corollary 6.17 *With the above notation we have:*

(i) $S_\beta(h)$ *is nonempty if and only if* β *is consistent.*

(ii) For any $h \in A$, $S_\beta(h)$ *is a closed (possibly empty, or singleton) interval contained in* [0, 1].

Proof (i) By Theorem 6.3, the extendability of β to a state is equivalent to the consistency of β.

(ii) If β is not extendable to a state we have nothing to prove. If β is extendable to a state, let us fix $j = 1, \ldots, m$. Then the nonempty set of states σ of A such that $\sigma(h_j) = \beta(h_j)$ is the intersection of the compact convex set $\mathsf{S}(A) \subseteq [0, 1]^A$ with the closed convex (a fortiori connected) set $S_{\beta\restriction h_j} \subseteq \mathsf{S}(A)$. As a consequence,

$$S_\beta = S_{\beta\restriction h_1} \cap \cdots \cap S_{\beta\restriction h_m} \subseteq \mathsf{S}(A).$$

Let the function $\varphi_h\colon \mathsf{S}(A) \to [0, 1]$ be defined by $\varphi_h(\sigma) = \sigma(h)$, for all $\sigma \in \mathsf{S}(A)$. By definition of the topology of $\mathsf{S}(A)$, φ_h is a continuous function. Its restriction $\varphi_h \restriction S_\beta = S_\beta(h)$ to the nonempty closed connected set S_β has a closed connected range. □

Historical and bibliographical notes. MV-algebraic states where introduced in [13], where also their basic properties were established.

Goodearl's monograph [8] provides a comprehensive account of states in unital ℓ-groups, and their relations with states in the C*-algebras of quantum systems. A proof of the Krein Milman theorem can be found in this book. See [8, 5.17, p. 85].

The categorical equivalence Γ was introduced in [12]. It is the key tool to show that MV-algebraic states have the same structural properties of the states of unital

ℓ-groups. As is always the case for categorical equivalences, (see, e.g., [11, IV, §4, Theorem 1]) there is an equivalence Ξ in the opposite direction of Γ. See TheBook, Sect. 7 for details on Ξ. While the archimedean property of the unit in unital ℓ-groups is not even definable in first-order logic, many structural properties of the *equational* class of MV-algebras carry over to unital ℓ-groups via Ξ. Examples are given by free, finitely presented and projective objects with their rational polyhedral spectral theory and algorithms.

The proof of the MV-algebraic consistency Theorem 6.3 given here is a simplification of the proofs in [15] and in [10].

For a strong form of Corollary 6.16 see [14, Theorem 19.8(i)].

Working in the context of Łukasiewicz logic and MV-algebras, in their paper [5], Fedel, Keimel, Montagna and Roth introduce a consistency criterion for *non-reversible bookmaking*. The "imprecise probabilities" introduced in that paper can be defined in terms of compact convex sets of probabilities or, equivalently, in terms of suitable sublinear functionals. The resulting consistency condition arises from these functionals just as the consistency notion considered in this paper arises from states. Also see [3].

De Finetti's characterization has been extended to (events coded by) formulas in various nonclassical logics. See, e.g., the work of Paris [17] for several modal logics, and his proofs of [17, Theorem 2 and Corollary 3]. In the paper [1] de Finetti's consistency criterion is extended to Gödel logic [9], a logic with a *discontinuous* implication connective. MV-algebraic states also have a fundamental importance in the study of non-boolean extensions of Carathéodory algebraic measure theory (see [16], and in particular, [18]).

In [7] one can find an algebraic characterization of de Finetti's criterion using a class of structures where MV-algebraic states are defined as internal operators. A stronger notion of consistency on MV-algebras is characterized in [4], also see [6].

References

1. Aguzzoli, S., Gerla, B., & Marra, V. (2008). De Finetti's No-Dutch-Book Criterion for Gödel logic. Special issue on "Many-valued logic and cognition." *Studia Logica, 90*, 25–41.
2. Boole, G. (1854). An investigation of the laws of thought. Walton and Maberley (p. 1958) Reprint.
3. Fedel, M., Hosni, H., & Montagna, F. (2011). A logical characterization of coherence for imprecise probabilities. *International Journal of Approximate Reasoning, 52*(8), 1147–1170.
4. Fedel, M., Hosni, H., & Montagna, F. (2018). Strict coherence on many-valued events. *The Journal of Symbolic Logic, 83*(1), 55–69.
5. Fedel, M., Keimel, K., Montagna, F., & Roth, W. (2013). Imprecise probabilities, bets and functional analytic methods in Łukasiewicz logic. *Forum Mathematicum, 25*(2), 405–441.
6. Flaminio, T., Godo, L., & Hosni, H. (2015). Coherence in the aggregate: A betting method for belief functions on many-valued events. *International Journal of Approximate Reasoning, 58*, 71–86.
7. Flaminio, T., & Montagna, F. (2007). An algebraic approach to states on MV-algebras. In M. Štěpnička, et al. (Eds.), Proceedings 5th EUSFLAT07 Conference, Ostrava (Czech Republic) (Vol. 2, pp. 201–206).

8. Goodearl, K. R. (1986). *Partially ordered abelian groups with interpolation*. American Mathematical Society.
9. Hájek, P. (1998). Metamathematics of fuzzy logic, Trend in Logic, Vol. 4. Kluwer.
10. Kühr, J., & Mundici, D. (2007). De Finetti theorem and Borel states in [0,1]-valued algebraic logic. *International Journal of Approximate Reasoning, 46*, 605–616.
11. Mac Lane, S. (2013). Categories for the working mathematician, Graduate Texts in Mathematics (Vol. 5, 2nd ed.). Springer.
12. Mundici, D. (1986). Interpretation of AF C^*-algebras in Łukasiewicz sentential calculus. *Journal of Functional Analysis, 65*, 15–63.
13. Mundici, D. (1995). Averaging the truth value in Łukasiewicz sentential logic. *Studia Logica, Special issue in honor of Helena Rasiowa, 55*, 113–127.
14. Mundici, D. (2011). Advanced Łukasiewicz calculus and MV-algebras. Trends in Logic (Vol. 35). Springer.
15. Mundici, D. (2019). Betting on continuous independent events. *Soft Computing, 23*, 2289–2295.
16. Pap, E. (Ed.). (2002). *Handbook of measure theory, I, II*. Amsterdam: North-Holland.
17. Paris, J. (2001). A note on the Dutch Book method. In G. De Cooman, T. Fine & T. Seidenfeld (Eds.), Proceedings of the Second International Symposium on Imprecise Probabilities and their Applications, ISIPTA 2001, Shaker Publishing Company, Ithaca, NY, USA (pp. 301–306). http://www.stats.org.uk/why-probability/Dutch-book/Paris2002.pdf.
18. Riečan, B., & Mundici, D. Probability on MV-algebras. In [16] (Vol. II, pp. 869–909).

Chapter 7
MV-Algebras and Kolmogorov's Triples $(\Omega, \mathcal{B}(\Omega), P)$

One can easily construct an affine homeomorphism of the state space of any MV-algebra A onto the state space of the quotient of A by its radical. This quotient is always a semisimple MV-algebra. Therefore, all relevant properties of states of an MV-algebra A still hold in the semisimple quotient $A/\mathsf{Rad}(A)$. Letting A range over MV-algebras, $A/\mathsf{Rad}(A)$ yields an isomorphic copy of every semisimple MV-algebra. By Theorem 7.4, any semisimple MV-algebra A can be represented as an algebra of continuous $[0, 1]$-valued functions defined on a compact Hausdorff space X. A homeomorphic copy of X is provided both by the space of maximal ideals of A with the hull-kernel topology, and by the space $\mathsf{hom}(A)$ with the topology of $[0, 1]^A$ inherited by restriction (Theorem 7.11). Theorem 7.13 yields an affine isomorphism κ of the state space of any (semisimple) MV algebra A onto the space of regular Borel probability measures on the maximal spectral space of A. As a consequence, in Theorem 7.18 we extend to MV-algebras the boolean product law, showing that product is the only consistency preserving continuous operation $\diamond\colon\ [0, 1]^2 \to [0, 1]$ for conjunctions of independent continuous $[0, 1]$-valued events. As another consequence, in Theorem 7.20 we prove that pairs (A, σ) with A an MV-algebra A and σ a state of A yield all Kolmogorov triples $(\Omega, \mathcal{B}(\Omega), P)$, for any compact Hausdorff space Ω, by letting $\Omega = \mu(A),\quad \mathcal{B}(\Omega) = \mathcal{B}(\mu(A)),\quad P = \kappa(\sigma)$.

7.1 The Maximal Spectral Space $\mu(A)$ of an MV-Algebra A

For any MV-algebra A we let $\mu(A)$ denote the space of maximal ideals of A endowed with the *hull-kernel topology*. A basis of closed sets for this topology is given by all sets of the form

$$\mathsf{F}_a = \{\mathfrak{m} \in \mu(A) \mid a \in \mathfrak{m}\}, \quad \text{letting } a \text{ range over the set of elements of } A. \tag{7.1}$$

D. Mundici, *Algebraic Probabilistic Consistency*, Trends in Logic 69,
https://doi.org/10.1007/978-3-031-98337-5_7

We say that $\mu(A)$ is the *maximal spectral space of* A. When A is a boolean algebra, $\mu(A)$ coincides with the Stone space of A.

Recall that all MV-algebras in this book are proper, $0 \neq 1$.

Proposition 7.1 *For any MV-algebra A,* $\mu(A)$ *is a nonempty compact Hausdorff space.*

Proof Let $\mathfrak{m} \subseteq A$ be maximal for the following conditions, for all $a, b, c, d \in A$:

$$(1 \notin \mathfrak{m}), \quad (a, b \in \mathfrak{m} \Rightarrow a \oplus b \in \mathfrak{m}), \quad (c \leq d \in \mathfrak{m} \Rightarrow c \in \mathfrak{m}).$$

Since the zero and the unit elements of A are distinct, the Axiom of Choice ensures the existence of at least one such subset $\mathfrak{m}$ of A, because the singleton zero satisfies these conditions, and so does the union of any increasing family of subsets of A satisfying these conditions. By direct inspection, $\mathfrak{m}$ is a maximal ideal of A. Furthermore, $\mu(A)$ has the Hausdorff separation property, since any two distinct elements of $\mu(A)$ are incomparable.

There remains to prove that the maximal spectral space of A is compact. To this purpose, again using the Axiom of Choice, let $\mathcal{I}$ be a maximal family of closed sets in $\mu(A)$ such that every finite subset of $\mathcal{I}$ has a nonempty intersection. We now claim that $\bigcap \mathcal{I} \neq \emptyset$. As a matter of fact, the set $\mathfrak{q} = \{a \in A \mid \mathsf{F}_a \in \mathcal{I}\}$ is a maximal ideal of A and $\bigcap \mathcal{I} = \{\mathfrak{q}\}$. We conclude that $\mu(A)$ is compact. □

Letting A range over all MV-algebras, in Theorem 7.4(iii) we will show that, up to homeomorphism, $\mu(A)$ yields every nonempty compact Hausdorff space.

Let $X \neq \emptyset$ be a set and $A \subseteq [0, 1]^X$ an MV-algebra of $[0, 1]$-valued functions on X, with the pointwise operations of $[0, 1]$. We say that A *separates points* (or, A is *separating*) if for any two distinct points $x, y \in X$ there is $f \in A$ such that $f(x) \neq f(y)$.

Definition 7.2 For any compact Hausdorff space $X \neq \emptyset$ we denote by $C(X, [0, 1])$ the MV-algebra of all continuous $[0, 1]$-valued functions on X equipped with the pointwise operations of the standard MV-algebra $[0, 1]$. Since every compact Hausdorff space is normal, $C(X, [0, 1])$ is separating.

Definition 7.3 The *radical* $\mathsf{Rad}(A)$ of an MV-algebra A is the intersection of its maximal ideals. A is said to be *semisimple* if $\mathsf{Rad}(A) = \{0\}$.

Notational and terminological conventions. The *zeroset* Zf of a function $f : X \to \mathbb{R}$ is defined by $Zf = f^{-1}(0) = \{x \in X \mid f(x) = 0\}$. More generally, for any subset $J \subseteq \mathbb{R}^X$ the *zeroset* ZJ of j is defined by $ZJ = \bigcap\{Zf \mid f \in J\}$. (This set is denoted V_J in TheBook, p.65.) When ZJ is a singleton we write $\dot{Z}J$ shorthand for the only element of ZJ.

Generalizing Theorem 4.6, we have:

Theorem 7.4 (Characterization of semisimple MV-algebras)
(i) An MV-algebra A is semisimple if and only if it is isomorphic to a separating MV-algebra of $[0, 1]$*-valued continuous functions defined on some (always) compact Hausdorff space X, in symbols,*

$$A \subseteq C(X, [0, 1]).$$

Thus in particular, every boolean algebra is a semisimple MV-algebra.
(ii) Let A be a separating MV-algebra of $[0, 1]$*-valued continuous functions defined on a compact Hausdorff space X. Then the map*

$$x \in X \mapsto \mathfrak{m}_x = \{f \in A \mid f(x) = 0\}$$

is a bijection of X onto $\boldsymbol{\mu}(A)$*. Its inverse is the map* $\dot{Z}$ *sending every maximal ideal* $\mathfrak{m} \in \boldsymbol{\mu}(A)$ *to the only member* $x = x_{\mathfrak{m}} \in X$ *of the set* $\dot{Z}\mathfrak{m} = \bigcap\{Zf \mid f \in \mathfrak{m}\}$.
(iii) Letting A range over all MV-algebras, $\boldsymbol{\mu}(A)$ *yields a homeomorphic copy of any possible nonempty compact Hausdorff space.*
(iv) Any two separating MV-algebras A, A' *of* $[0, 1]$*-valued continuous functions on the same compact Hausdorff space have homeomorphic maximal ideal spaces.*

Proof (i) See TheBook, Corollary 3.6.8. The semisimplicity of any boolean algebra now immediately follows from Stone duality. (ii) See TheBook, 3.4.3(i). (iii)–(iv) Immediate from (ii). □

Strengthening (ii), Theorem 7.11 will show that the bijection $x \mapsto \mathfrak{m}_x$ is a homeomorphism of X onto $\boldsymbol{\mu}(A)$ for any MV-algebra $A \subseteq C(X, [0, 1])$.

Proposition 7.5 *Let A be an MV-algebra.*
(i) For any state σ *of A and* $\epsilon \in \mathsf{Rad}(A)$, $\sigma(\epsilon) = 0$.

(ii) The function $\rho\colon \sigma \in \mathsf{S}(A) \mapsto \sigma' \in \mathsf{S}(A/\mathsf{Rad}(A))$ *defined by the stipulation* $\sigma(a) = \sigma'(a/\mathsf{Rad}(A))$, *(for all* $a \in A$*) is an affine homeomorphism of the state space of A onto the state space of* $(A/\mathsf{Rad}(A))$*. In symbols,*

$$\rho\colon \mathsf{S}(A) \cong \mathsf{S}(A/\mathsf{Rad}(A)).$$

Proof (i) With the notation of TheBook (2.7), for $n = 1, 2, \ldots$ let

$$n.x \quad \text{be shorthand for} \quad \underbrace{x \oplus x \cdots \oplus x}_{n \text{ times}}. \tag{7.2}$$

Following Definition 3.6.3 in TheBook, an *infinitesimal* of A is a nonzero element ϵ of A satisfying $n.\epsilon \leq \neg\epsilon$ for all $n = 1, 2, \ldots$. It follows that $n.\epsilon < \neg\epsilon$ for all $n =$

$1, 2, \ldots,$ since $n.\epsilon = \neg\epsilon$ would imply $2n.\epsilon = 1 > \neg\epsilon$ (as $\epsilon > 0$), which is impossible. As proved in TheBook, 3.6.4, $\mathsf{Rad}(A)$ consists of the zero element of A together with all infinitesimals of A. By Proposition 6.9(iii), if $\epsilon = 0$ we are done. By way of contradiction, assume $\sigma(\epsilon) \neq 0$ for some infinitesimal ϵ. Then $\sigma(\epsilon)$ is a real number > 0. From the additivity property of σ it follows that for all $x \in [0, 1]$ there is $m = 1, 2, \ldots$ such that $\sigma(m.\epsilon) \geq x$. In particular, $1 = \sigma(m.\epsilon) \geq \sigma(\neg\epsilon)$ for all suitably large m. The monotonicity property of σ (Proposition 6.9(i)) yields the inequality $\sigma(n.\epsilon) \leq \sigma(\neg\epsilon)$ for all $n = 1, 2, \ldots,$ and hence, for all large m, $1 = \sigma(m.\epsilon) = \sigma(\neg\epsilon)$. By Proposition 6.9(iv), $\sigma(\epsilon) = 0$, a contradiction.

(ii) Immediate from (i). □

Proposition 7.6 *Let A be an MV-algebra.*

(i) For any maximal ideal $\mathfrak{m}$ of A there is a unique pair $(\mathfrak{i}_{\mathfrak{m}}, I_{\mathfrak{m}})$, with $I_{\mathfrak{m}}$ an MV-subalgebra of the standard MV-algebra $[0, 1]$, and $\mathfrak{i}_{\mathfrak{m}}$ an isomorphism of $A/\mathfrak{m}$ onto $I_{\mathfrak{m}}$. Letting the map $\mathfrak{H}\colon \boldsymbol{\mu}(A) \to \mathsf{hom}(A)$ be defined by $(\mathfrak{H}(\mathfrak{m}))(a) = \mathfrak{i}_{\mathfrak{m}}\left(\frac{a}{\mathfrak{m}}\right)$ for all $a \in A$, we have

$$\ker \mathfrak{H}(\mathfrak{m}) = \mathfrak{m} \textit{ and } \mathfrak{H} \textit{ is injective.} \tag{7.3}$$

(ii) Any $\eta \in \mathsf{hom}(A)$ has the form $\eta = \mathfrak{H}(\mathfrak{m})$ for the uniquely determined $\mathfrak{m} \in \boldsymbol{\mu}(A)$ given by $\mathfrak{m} = \ker \eta$. Hence $\mathfrak{H}$ is a bijection of $\boldsymbol{\mu}(A)$ onto $\mathsf{hom}(A)$. The inverse of $\mathfrak{H}$ is the map $\ker\colon \mathsf{hom}(A) \to \boldsymbol{\mu}(A)$. Thus for any $\zeta \in \mathsf{hom}(A)$ and $a \in A$, from (7.3) we obtain

$$\mathfrak{H}(\ker \zeta) = \zeta \quad \textit{and} \quad \eta(a) = \mathfrak{i}_{\ker \eta}\left(\frac{a}{\ker \eta}\right). \tag{7.4}$$

Proof (i) Since $\mathfrak{m}$ is maximal, by TheBook 1.2.10 the quotient MV-algebra $A/\mathfrak{m}$ is simple, i.e., $\{0\}$ is the only (proper, maximal) ideal of $A/\mathfrak{m}$. By TheBook 3.5.1, $A/\mathfrak{m}$ is isomorphic to a subalgebra I of $[0, 1]$. If $A/\mathfrak{m}$ is isomorphic to another subalgebra I' of $[0, 1]$ then $I \cong I'$ whence by TheBook 7.2.6, $I = I'$. Thus the MV-algebra $I = I_{\mathfrak{m}}$ is uniquely determined by $\mathfrak{m}$. Likewise, the isomorphism $\mathfrak{i}_{\mathfrak{m}}\colon A \cong A/\mathfrak{m}$ is uniquely determined by $\mathfrak{m}$. Trivially, $\mathfrak{H}(\mathfrak{m}) \in \mathsf{hom}(A)$. Furthermore, the range of $\mathfrak{H}(\mathfrak{m})$ is a simple MV-algebra. By TheBook 1.2.10 and 3.5.1, $\ker \mathfrak{H}(\mathfrak{m})$ is a maximal ideal of A. Specifically, for all $a \in A$ we can write

$$a \in \mathfrak{m} \Leftrightarrow a/\mathfrak{m} = 0 \in A \Leftrightarrow \mathfrak{i}_{\mathfrak{m}}(a/\mathfrak{m}) = (\mathfrak{H}(\mathfrak{m}))(a) = 0 \in [0, 1] \Leftrightarrow a \in \ker \mathfrak{H}(\mathfrak{m}),$$

which settles (7.3) and immediately yields the injectivity of $\mathfrak{H}$.

(ii) As already mentioned, for every $\eta \in \mathsf{hom}(A)$ its kernel is a maximal ideal of A. From $\ker \eta$ we obtain the homomorphism $\mathfrak{H}(\ker \eta)$ of A into $[0, 1]$. By (7.3),

$$\ker(\mathfrak{H}(\ker \eta)) = \ker \eta. \tag{7.5}$$

Claim: $\mathfrak{H}(\mathsf{ker}\,\eta) = \eta$.

We have to prove that different homomorphisms ζ, ζ' of A into $[0, 1]$ have different kernels. We will argue by an MV-algebraic simplification of Hölder's time-honored characterization of the additive ℓ-group of real numbers with unit 1.

By hypothesis, there is $a \in A$ with $\zeta(a) \neq \zeta'(a)$, say without loss of generality $\zeta(a) < \zeta'(a)$.

Case 0. If $\zeta(a) = 0$ we are done.

Case 1. If $\zeta'(a) = 1$, letting $a_1 = \neg a$ we revert to Case 0 (with a_1 in place of a), and we are done.

Case 2. If $\zeta(a) < 1/2 \leq \zeta'(a)$, letting $a_2 = a \oplus a$ we have $\zeta(a_2) < 1$, $\zeta'(a_2) = 1$, and we are in Case 1 with a_2 in place of a. One argues similarly if $\zeta'(a) < 1/2 \leq \zeta(a)$.

...

Case t. Having inductively constructed $a_t \in A$, a judicious choice of

$$a_{t+1} \in \{\neg a_t, a_t \oplus a_t\}$$

will either bring us to a case of type 0,1,2 (whence we are done), or else will *result in doubling the distance between* $\zeta(a_{t+1})$ and $\zeta'(a_{t+1})$.

Proceeding inductively, after a finite number t^* of doubling steps, $\zeta(a_{t^*})$ and $\zeta'(a_{t^*})$ will be separated by the midpoint $1/2$, as in Case 2. Then, as a final step, replacing a_{t^*} by $a_{t^*} \oplus a_{t^*}$ we revert to Case 1, and we may conclude that ζ and ζ' have different kernels, as desired.

We have proved that the map $\mathsf{ker}\colon \mathsf{hom}(A) \to \boldsymbol{\mu}(A)$ is a bijection onto $\boldsymbol{\mu}(A)$ whose inverse is $\mathfrak{H}$. □

Proposition 7.7 *Let A be a separating MV-subalgebra of $C(X, [0, 1])$ for some compact Hausdorff space X. Then the composite map $\dot{Z}\,\mathsf{ker}\colon \eta \in \mathsf{hom}(A) \mapsto \dot{Z}\,\mathsf{ker}\,\eta$ is a bijection of $\mathsf{hom}(A)$ onto X. Its inverse is the map*

$$\mathsf{eval} = (\dot{Z}\,\mathsf{ker})^{-1}\colon X \to \mathsf{hom}(A)$$

sending any $x \in X$ to the only $\eta = \eta_x \in \mathsf{hom}(A)$ whose kernel satisfies $\dot{Z}\mathsf{ker}\,\eta = x$. In other words, $\mathsf{eval}(x)$ is the homomorphism η_x of A into $[0, 1]$ that evaluates *every function in $f \in C(X, [0, 1])$ at the point x,*

$$(\mathsf{eval}(x))(f) = f(x). \tag{7.6}$$

For every $x \in X$ and $\mathfrak{m} \in \boldsymbol{\mu}(A)$ we have

$$\eta_x(f) = f(x) \ \ \textit{and} \ \ \eta_{x_\mathfrak{m}}(f) = \mathfrak{i}(f/\mathfrak{m}) = (\mathfrak{H}(\mathfrak{m}))(f) = f(x_\mathfrak{m}). \tag{7.7}$$

Proof Combine Theorem 7.4 with Proposition 7.6. □

7.2 McNaughton's Theorem, Free and Semisimple MV-Algebras

Since MV-algebras are defined by equations, every MV-algebra is the quotient of some free MV-algebra. Free MV-algebras are an important class of semisimple MV-algebras, with a concrete geometric realization.

Throughout this book the adjective "linear" is understood in the affine sense. For $k = 1, 2, \ldots$, a *McNaughton function* $f\colon [0, 1]^k \to [0, 1]$ is a continuous piece-wise linear function consisting of finitely many pieces, each piece being represented by a linear polynomial with integer coefficients. For κ an infinite cardinal, following TheBook, 3.1.7, we define a *McNaughton function on* $[0, 1]^\kappa$ as the cylindrification $f_{\uparrow\kappa}$ of a McNaughton function $f(x_{\alpha_1}, \ldots, x_{\alpha_t}) \in [0, 1]^t$, for ordinals $\alpha_1, \ldots, \alpha_t < \kappa$. In other words,

$$f_{\uparrow\kappa}(x) = f(x_{\alpha_1}, \ldots, x_{\alpha_t}) \in [0, 1]^t, \text{ for all } x \in [0, 1]^\kappa .$$

The simplest possible examples of McNaughton functions, are the constants 0 and 1. Also, for any integer $k = 1, 2, \ldots$ let $\pi_1, \ldots, \pi_k$ be the *projection* (also known as identity, or coordinate) functions on $[0, 1]^k$. For each $i = 1, \ldots, k$ and $x \in [0, 1]^k$ the function π_i is defined by $\pi_i(x) = x_i$.

Theorem 7.8 (McNaughton) *For any $k = 1, 2, \ldots,$ the free MV-algebra over the free generating set $\{\pi_1, \ldots, \pi_k\}$ is (isomorphic to) the MV-algebra $\mathcal{M}([0, 1]^k)$ of all k-variable McNaughton functions equipped with the pointwise operations $\neg x = 1 - x$ and $x \oplus y = \min(1, x + y)$ of the standard MV-algebra* $[0, 1]$. *For κ an infinite cardinal, the free MV-algebra over κ generators is similarly defined as the MV-algebra $\mathcal{M}([0, 1]^\kappa)$ of all McNaughton functions over* $[0, 1]^\kappa$.

Proof A constructive proof of the original McNaughton theorem can be found in TheBook §9.1. □

Theorem 7.9 *Let κ be a cardinal $\neq 0$. Then for any semisimple MV-algebra A with κ generators there is a nonempty compact Hausdorff set $X \subseteq [0, 1]^\kappa$ such that A is isomorphic to the MV-algebra $\mathcal{M}([0, 1]^\kappa) \restriction X$ of restrictions to X of the McNaughton functions of $\mathcal{M}([0, 1]^\kappa)$. $\mathcal{M}([0, 1]^\kappa) \restriction X$ automatically separates points of X.*

Proof TheBook, Theorem 3.6.7. □

Proposition 7.10 *Fix $\kappa \neq 0$ a finite or infinite cardinal, let O be an arbitrarily chosen open neighborhood in $[0, 1]^\kappa$ of a point x. Then some function $f \in \mathcal{M}([0, 1]^\kappa)$ vanishes over the closed set $[0, 1]^\kappa \setminus O$ and is constantly equal to 1 over an open neighborhood of x in* $[0, 1]^\kappa$.

Proof It is enough to argue for an infinite κ. We first assume that x lies in the interior of $[0, 1]^\kappa$. By definition of the product topology of $[0, 1]^\kappa$, up to an inessential permutation of a finite number n of ordinals preceding κ, we may safely assume that O contains the open rectangle $R \ni x$ given by

$$R = \{y \in [0,1]^\kappa \mid 0 < p_i/q_i < y_i < r_i/s_i < 1 \quad \text{for all} \quad i = 1, \ldots, n\},$$

for suitable rationals $0 < p_i/q_i < x_i < r_i/s_i < 1$. For each $i = 1, \ldots, n$, using the techniques of TheBook §3.2, one easily constructs a one-variable McNaughton function $f_i = f_i(x_i)$ satisfying the following conditions: f_i constantly vanishes over $[0, 1] \setminus [p_i/q_i, r_i/s_i]$, and f_i takes the constant value 1 over an open interval containing x.
Then the cylindrification of the function $f = f_1(x_i) \vee \cdots \vee f_n(x_n)$ has the desired properties. The proof when x lies on the boundary of $[0, 1]^\kappa$ follows from an inessential modification of the above proof. □

Strengthening Theorem 7.4, for any semisimple MV-algebra A, Theorem 7.9 provides the isomorphism

$$A \cong \mathcal{M}([0,1]^\kappa) \restriction X$$

for some cardinal $\kappa > 0$ and compact Hausdorff space $X \subseteq [0, 1]^\kappa$. As shown by the following result, this representation is of great help to visualize the maximal spectral space $\boldsymbol{\mu}(A)$ and the set of homomorphisms of A into $[0, 1]$.

Theorem 7.11 *Let A be a semisimple MV-algebra. Then:*

$\dot{Z}$ *is a homeomorphism of $\boldsymbol{\mu}(A)$ onto X.*

ker *is a homeomorphism of* hom(A) *onto $\boldsymbol{\mu}(A)$.*

Proof By Theorem 7.9, for some compact Hausdorff subspace of $[0, 1]^\kappa$, A may be identified with $\mathcal{M}([0, 1]^\kappa) \restriction X$. It is sufficient to argue assuming $A = \mathcal{M}([0, 1]^\kappa)$, with κ an infinite cardinal. (The proof for $A = \mathcal{M}([0, 1]^\kappa) \restriction X$ then follows as a routine exercise.) In view of the bijections of Theorem 7.4(ii) and Proposition 7.6(ii) there remains to prove that these bijections are continuous.

- *From an open in $[0, 1]^\kappa$ to an open in $\boldsymbol{\mu}(A)$ via the map $x \mapsto \mathfrak{m}_x$.* Let $I^\uparrow$ be a subbasic open in $[0, 1]^\kappa$ given by the cylindrification in $[0, 1]^\kappa$ of an open segment $I = I_\alpha$ with rational endpoints (for some $\alpha < \kappa$). Following TheBook §3.2, we will construct a function $t = t_I \in \mathcal{M}([0, 1])$ of the variable x_α whose *open support*

  $$\{y \in [0,1]_\alpha \mid x_\alpha(y) > 0\}$$

  coincides with I in the αth copy of the unit interval $[0, 1]_\alpha$. Cylindrification transforms t into a function in $\mathcal{M}([0, 1]^\kappa)$ only depending on the variable x_α. The set

  $$\mathsf{F}_t = \{\mathfrak{n} \in \boldsymbol{\mu}(\mathcal{M}([0,1]^\kappa)) \mid t \notin \mathfrak{n}\}$$

  is an open (basic) set of $\boldsymbol{\mu}(\mathcal{M}([0, 1]^\kappa))$. Now note that $\{\dot{Z}\mathfrak{n} \mid \mathfrak{n} \in \mathsf{F}_t\}$ coincides with $I^\uparrow \subseteq [0, 1]^\kappa$.
- *From any open set in $\boldsymbol{\mu}(A)$ to an open set in* hom(A) *via the map $\mathfrak{H}$.* We may restrct to the case when U is a basic open of $\boldsymbol{\mu}(A)$ of the form

$$\mathsf{F}_f = \{\mathfrak{n} \in \boldsymbol{\mu}(A) \mid f \notin \mathfrak{n}\} \text{ for some } f \in \mathcal{M}([0,1]^\kappa).$$

The bijection of Proposition 7.6

$$\mathsf{ker}^{-1} = \mathfrak{H}\colon \boldsymbol{\mu}(\mathcal{M}([0,1]^\kappa)) \to \mathsf{hom}(\mathcal{M}([0,1]^\kappa))$$

maps F_f onto the set $\mathfrak{H}_{\mathsf{F}_f}$ of homomorphisms of A into $[0, 1]$ given by

$$\{\mathfrak{H}(\mathfrak{m}) \mid f \notin \mathfrak{m}\} = \{\eta \in \mathsf{hom}(\mathcal{M}([0,1]^\kappa)) \mid \eta = \mathfrak{H}(\mathfrak{m}) \text{ for some } \mathfrak{m} \text{ with } f \notin \mathfrak{m}\}.$$

With $\mathsf{quot}_\mathfrak{m}$ denoting the homomorphism of A onto its quotient $A/\mathfrak{m}$, by definition of $\mathfrak{H}$ and Proposition 7.6 we can write:

$$\begin{aligned}\{\mathfrak{H}(\mathfrak{m}) \mid f \notin \mathfrak{m}\} &= \{\eta \mid \eta = \mathfrak{i}_\mathfrak{m}(\mathsf{quot}_\mathfrak{m}) \text{ for some } \mathfrak{m} \text{ with } f/\mathfrak{m} \neq 0\}\\ &= \{\eta \mid \text{the only } \mathfrak{m} \text{ such that } \eta = \mathfrak{i}_\mathfrak{m}(\mathsf{quot}_\mathfrak{m}) \text{ satisfies } f/\mathfrak{m} \neq 0\}\\ &= \{\eta \mid f/\mathsf{ker}(\eta) \neq 0 \neq \mathfrak{i}_{\mathsf{ker}(\eta)} f/\mathsf{ker}(\eta)\}\\ &= \{\eta \mid \eta(f) > 0\}, \text{ an open set in } \mathsf{hom}(A).\end{aligned}$$

The proof is complete. □

Corollary 7.12 *Any two separating MV-algebras A, A' of continuous* $[0, 1]$*-valued functions on homeomorphic compact Hausdorff spaces $X \cong X'$ have:*

(i) Homeomorphic maximal spectral spaces $\boldsymbol{\mu}(A) \cong \boldsymbol{\mu}(A') \cong X$.
(ii) Homeomorphic spaces $\mathsf{hom}(A) \cong \mathsf{hom}(A')$ *of homomorphisms into* $[0, 1]$.
(iii) Homeomorphic extremal state spaces $\mathsf{ext}\,\mathsf{S}(A) \cong \mathsf{ext}\,\mathsf{S}(A') \cong \mathsf{hom}(A)$.
(iv) Affinely homeomorphic compact Hausdorff spaces $\mathsf{S}(A) \cong \mathsf{S}(A') \cong \mathsf{cl\,conv\,hom}(A) \cong \mathsf{cl\,conv\,ext\,S}(A)$.

Proof Combine Corollary 6.11 with Theorems 6.12 and 7.11. □

7.3 Kroupa's Theorem: $\mathsf{S}(A) \cong \mathcal{P}(\mu(A))$

For any compact Hausdorff space X we denote by $\mathcal{P}(X)$ the convex set of regular Borel probability measures ν on X, where ν is said to be *regular* if for all Borel sets $Y \subseteq X$,

$$\nu(Y) = \sup\{\nu(K) \mid K \subseteq Y,\ K \text{ closed}\} = \inf\{\nu(U) \mid U \supseteq Y,\ U \text{ open}\}. \tag{7.8}$$

A *real vector lattice* (also known as a *Riesz space*) is a real vector space V which is also a lattice by a partial order relation $x \leq y$ having the property that for all $x, y, t \in V$ and $0 \leq \gamma \in \mathbb{R}$, $x \leq y \Rightarrow x + t \leq y + t$ and $\gamma x \leq \gamma y$. The definition of a *rational vector lattice* is the same, with V a rational vector space. V is *unital* if it is equipped with a distinguished (strong order) *unit* u, i.e., an element whose positive integer multiples eventually dominate each element of V.

By Proposition 7.5, any MV-algebra has the same states and the same maximal spectral spaces as the quotient by its radical. In view of this proposition, restriction to semisimple MV-algebras entails no essential loss of generality when considering the main structural properties of MV-algebras and their states.

The following result extends to MV-algebras Carathéodory's extension Theorem 4.10:

Theorem 7.13 (Kroupa) *Let A be a semisimple MV-algebra. In view of Theorem 7.4 let us identify A with a separating MV-subalgebra of $C(X, [0, 1])$, for X a compact Hausdorff space. It follows that a map $\sigma : A \to [0, 1]$ is a state if and only if there is a (necessarily unique) regular Borel probability measure κ_σ on X such that*

$$\sigma(f) = \int_X f(x)\, \mathrm{d}\kappa_\sigma(x) \quad \textit{for every } f \in C(X, [0, 1]). \tag{7.9}$$

Proof For the nontrivial direction, Theorem 7.11 yields a (canonical) homeomorphism of X onto the (automatically nonempty) maximal spectral space of A. Let G be the group of continuous functions on X generated by A in the real vector lattice $\mathbb{R}^X$ of all real valued functions on X. Since the constant function 1 over X is a unit of G we may write $\Gamma(G, 1) = A$.

Let $\mathsf{S}(G, 1)$ denote the set of states of $(G, 1)$. The unital *rational* vector lattice generated by $(G, 1)$ in $\mathbb{R}^X$ is known as the *divisible hull* $(\overline{G}, 1)$ of $(G, 1)$. There is a unique unital ℓ-embedding $(G, 1) \hookrightarrow (\overline{G}, 1)$. Any state τ of $(\overline{G}, 1)$ is *monotonic*: $a, b \in \overline{G} \Rightarrow \tau(f) \leq \tau(g)$. For any $r \in \mathbb{Q}$ the function constantly equal to r over X belongs to $\overline{G}$.

Let $C(X, \mathbb{R})$ denote the Banach algebra of real-valued continuous functions defined on X with the sup norm $||\cdot||_\infty$. Let $\mathsf{S}(C(X, \mathbb{R}))$ denote the set of normalized positive linear functionals on $C(X, \mathbb{R})$.

Claim 1: The restriction operation $\sigma \in \mathsf{S}(C(X, \mathbb{R})) \mapsto \sigma \upharpoonright (\overline{G}, 1)$ is an affine isomorphism of $\mathsf{S}(C(X, \mathbb{R}))$ onto $\mathsf{S}(\overline{G}, 1)$.

We must prove that every $\sigma \in \mathsf{S}(C(X, \mathbb{R}))$ is the unique extension to $C(X, \mathbb{R})$ of its restriction $\rho = \sigma \upharpoonright \overline{G}$. To this purpose let $h \in \overline{G}$ and $\nu \in \mathbb{R}_{\geq 0}$ be such that $||h||_\infty \leq \nu$ for some $\nu \geq 0$. Then we can write

$$\begin{aligned}
|\rho(h)| &= |\rho((h \vee 0) - (-h \vee 0))| \\
&= |\rho(h \vee 0) - \rho(-h \vee 0)| \\
&\leq \rho(h \vee 0) + \rho(-h \vee 0) \\
&\leq \rho(\nu 1), \quad \text{as } (h \vee 0) \wedge (-h \vee 0) = 0 \text{ and } \sup\{|h(x)| \mid x \in X\} \leq \nu \\
&= \nu, \quad \text{because } \rho(1) = 1 \text{ and } \rho \text{ is additive.}
\end{aligned}$$

In particular, for all $k, k' \in \overline{G}$ and $\epsilon > 0$,

$$||k - k'||_\infty < \epsilon \text{ implies } |\rho(k) - \rho(k')| = |\rho(k - k')| < \epsilon,$$

which ensures the continuity of ρ in the norm topology of the Banach algebra $C(X, \mathbb{R})$.

By the Stone-Weierstrass theorem, (lattice version)[1] the divisible ℓ-group $\overline{G}$ is norm-dense in $C(X, \mathbb{R})$. Since ρ is continuous, upon writing

$$\rho^*(\lim f_i) = \lim \rho(f_i), \quad (f_i \in \overline{G})$$

we can uniquely extend ρ to a *function* $\rho^* \colon C(X, \mathbb{R}) \to \mathbb{R}$. Trivially, ρ^* is linear and $\rho^*(1) = 1$. There remains to prove that ρ^* is positive. Let $0 \leq l \in C(X, \mathbb{R})$, with the intent of showing $\rho^*(l) \geq 0$. Since $\overline{G}$ is norm-dense in $C(X, \mathbb{R})$ there is a sequence $l_1, l_2, \ldots$ of elements of $\overline{G}$ such that $\lim l_i = l$. Hence, $\lim l_i \vee 0 = l \vee 0 = l$. (Note that also each function $l_i \vee 0$ belongs to $\overline{G}$, and $\rho(l_i \vee 0) \geq 0$, by definition of $\mathsf{S}(\overline{G}, 1)$.) As a consequence,

$$0 \leq \lim \rho(l_i \vee 0) = \rho^*(\lim(l_i \vee 0)) = \rho^*(l),$$

whence ρ^* is positive. We have shown that every member of $\mathsf{S}(C(X), \mathbb{R}))$ is the unique extension of its restriction to $\overline{G}$. The inverse of the extension map $\rho \mapsto \rho^*$ is the desired affine isomorphism of $\mathsf{S}(C(X, \mathbb{R}))$ onto $\mathsf{S}(\overline{G}, 1)$. Claim 1 is thus settled.

Next we observe that every state $\tau \in \mathsf{S}(G, 1)$ is uniquely extendable to a state $\bar{\tau} \in \mathsf{S}(\overline{G}, 1)$, upon setting $\bar{\tau}(r\, g) = r\, \tau(g)$ for any $g \in G$ and $r \in \mathbb{Q}$. All states of $(\overline{G}, 1)$ arise in this way from the states of $(G, 1)$. We can now immediately infer the following:

$$\text{restriction to } G \text{ is an affine isomorphism of } \mathsf{S}(\overline{G}, 1) \text{ onto } \mathsf{S}(G, 1). \tag{7.10}$$

Likewise, by Theorem 6.10,

$$\text{restriction to } A \text{is an affine isomorphism of } \mathsf{S}(G, 1) \text{ onto } \mathsf{S}(A). \tag{7.11}$$

To conclude the proof, let $\mathcal{P}(X)$ denote the convex set of regular Borel probability measures on X. The Riesz-Markov-Kakutani representation theorem[2] yields an affine isomorphism $\alpha \in \mathsf{S}(C(X, \mathbb{R})) \mapsto \nu_\alpha \in \mathcal{P}(X)$ such that

$$\alpha(f) = \int_X f(x)\, \mathrm{d}\nu_\alpha(x) \quad \text{for any} \quad f \in C(X, \mathbb{R}). \tag{7.12}$$

Let $\sim$ denote affine isomorphism. Claim 1 and (7.10)–(7.12) yield affine isomorphisms

[1] For a proof see the bibliographical notes for references at the end of this chapter.

[2] For a proof see the bibliographical notes.

$$\mathsf{S}(A) \sim \mathsf{S}(G, 1) \sim \mathsf{S}(\overline{G}, 1) \sim \mathsf{S}(C(X, \mathbb{R})) \sim \mathcal{P}(X). \tag{7.13}$$

Composition yields an affine isomorphism $\kappa \colon \mathsf{S}(A) \sim \mathcal{P}(X)$. By (7.12), κ transforms any state σ of A into a regular Borel probability measure $\kappa_\sigma \in \mathcal{P}(X)$ satisfying the identity

$$\sigma(f) = \int_X f(x)\, \mathrm{d}\kappa_\sigma(x) \quad \text{for any} \quad f \in C(X, \mathbb{R}).$$

From (7.13) and the inclusions $C(X, [0, 1]) \supseteq A \subseteq (G, 1) \subseteq (\overline{G}, 1) \subseteq C(X, \mathbb{R})$, it follows that κ_σ is uniquely determined by σ. □

The proof has shown:

Corollary 7.14 *For any MV-algebra A the map*

$$\kappa \colon \sigma \in \mathsf{S}(A) \mapsto \kappa(\sigma) = \kappa_\sigma \in \mathcal{P}(\boldsymbol{\mu}(A))$$

is an affine isomorphism of the state space $\mathsf{S}(A)$ *onto the space* $\mathcal{P}(\boldsymbol{\mu}(A))$ *of regular Borel probability measures on the maximal spectral space* $\boldsymbol{\mu}(A)$. □

Corollary 7.15 *With* $\mathcal{M}([0, 1])$ *the free one-generated MV-algebra, we have affine homeomorphisms*

$$\mathsf{S}(\mathcal{M}([0, 1])) \simeq \mathcal{P}(\boldsymbol{\mu}(\mathcal{M}([0, 1]))) \cong \mathcal{P}([0, 1]),$$

and homeomorphisms

$$\mathsf{ext}\,\mathcal{P}([0, 1]) \cong \mathsf{ext}\,\mathsf{S}(\mathcal{M}([0, 1])) \cong \mathsf{hom}(\mathcal{M}([0, 1])) \cong \boldsymbol{\mu}(\mathcal{M}([0, 1])) \cong [0, 1]\,.$$

Proof Combine Kroupa's theorem with Corollaries 7.12 and 7.14. □

7.4 The Product Law for Independent MV-Algebras

On page 168 of his 1905 lecture notes "Logische Prinzipien des mathematischen Denkens", Hilbert writes:

> Wir fassen das einfach als Definitionen auf, wiewohl im gegenwärtigen Zustande der Entwicklung besonders die Bezeichnungen "Axiom" und Definition noch etwas durcheinandergehen.
>
> (We simply take this as definitions, although in the present state of development, especially the terms "axiom" and definition are still a bit confused.)

In probability theory textbooks the additivity of probability for disjunctions of incompatible events is an *axiom*, while the "product law" for the probability for conjunctions of independent events is the *definition* of (stochastic) independence. In boolean logic one may express the incompatibility of events h, k by writing $h \wedge k = 0$, whereas there is no boolean connective to express the independence of h and k. Theorem 4.3 shows the uniqueness of product to preserve the consistency of a book of betting odds on conjunctions of algebraically independent boolean events. Our aim in this section is to extend this result to continuous $[0, 1]$-valued events in the framework of MV-algebras.

Boolean algebras (*qua* MV-algebras) are semisimple, while many interesting examples of MV-algebras have infinitesimals, which is an equivalent condition for failure of semisimplicity. However, by Proposition 7.5, the state space of any MV-algebra A can be identified with the state space of the semisimple MV-algebra $A/\mathsf{Rad}(A)$. As a consequence, our planned restriction to semisimple MV-algebras is inessential for the purposes of this chapter.

While boolean free product constructions had a major role in the proofs of the final theorems in Chapter 4, the appropriate tool to deal with algebraic independence in MV-algebras is given by tensor product, rather than MV-algebraic free product. By a quirk of fate, in boolean algebras (but not in MV-algebras) free products coincide with tensor products. Therefore, Theorem 7.18 below is a genuine extension of the boolean algebraic Theorem 4.3.

The semisimple tensor product of two MV-algebras. The semisimple tensor product construction is the key tool to deal with "independent" sets of continuous events $E \subseteq A$ and $F \subseteq B$ in semisimple MV-algebras A, B. By Theorem 7.4, A and B can be identified with separating MV-algebras of $[0, 1]$-valued continuous functions on compact Hausdorff spaces X and Y, respectively homeomorphic to $\boldsymbol{\mu}(A)$ and $\boldsymbol{\mu}(B)$. Once this is done, the *semisimple tensor product* $A \otimes B$ may be defined as the MV-algebra of $[0, 1]$-valued functions on the cartesian product $X \times Y$ generated by the functions of the form $f \cdot g$, for all $f \in A \subseteq C(X, [0, 1])$ and $g \in B \subseteq C(Y, [0, 1])$. The MV-algebraic embeddings

$$\mathsf{cyl}\uparrow : A \hookrightarrow A \otimes B \quad \text{and} \quad \mathsf{cyl}_{\rightarrow} : B \hookrightarrow A \otimes B$$

transform any $f \in A$ and $g \in B$ into their cylindrifications defined by

$$f_{\mathsf{cyl}\uparrow}(x, y) = f(x) \text{ and } g_{\mathsf{cyl}_{\rightarrow}}(x, y) = g(y), \text{ for all } x \in X \text{ and } y \in Y. \tag{7.14}$$

With this approach to the semisimple tensor product, from Corollary 4.8 it follows that the tensor product of two boolean algebras, *qua* semisimple MV-algebras, is the same as their free product. Furthermore, this approach dispenses with all technicalities concerning "universal bimorphisms", which are usually found in the literature on general MV-algebraic tensor products.

Definition 7.16 The canonical injections $A \hookrightarrow A \otimes B \hookleftarrow B$ send A and B onto subalgebras A', B' of the (always semisimple) tensor product $A \otimes B$. The MV-algebras A' and B' are said to be *independent subalgebras* of $A \otimes B$, and are conveniently identified with A and B. Correspondingly, any two finite subsets $E \subseteq A$ and $F \subseteq B$ are said to be *independent*.

We also say that A and B are *independent subalgebras of* C, whenever C is an MV-algebra containing $A \otimes B$ as a subalgebra.

Any function on $X \times Y$ of the form $f_{\text{cyl}\uparrow} \cdot g_{\text{cyl}_\rightarrow}$ $(f \in A = A',\ g \in B = B')$ is said to be a *pure tensor*.

Remark 7.17 The MV-algebra $A \otimes B$ is generated by its pure tensors. For each $a \in A$ and $b \in B$ the $[0, 1]$-valued event a "and" b is algebraically defined as the pure tensor $a \otimes b$. Trivially, if a and b are nonzero, then so is $a \otimes b$.

Now assume A, B, C to be boolean algebras. Then Definitions 7.16 and 4.1 agree, $A \otimes B$ is generated by its elements of the form $a \wedge b$, $(a \in A,\ \ b \in B)$ and coincides with the free product $A \amalg B$. In more detail, let us identify A and B with the boolean algebras of all continuous $\{0, 1\}$-valued functions on their Stone spaces $\bar{A}$ and $\bar{B}$. Then cylindrification identifies A and B with independent boolean algebras of continuous functions on $\bar{A} \times \bar{B}$, depending on disjoint sets of variables. It is easy to see that the pure tensor $a \otimes b$ coincides with the product $a \cdot b$ of the continuous boolean functions a, b, and also coincides with $a \wedge b$ and with $a \odot b$.

This no longer holds in general for MV-algebras. The pure tensor $a \otimes b$ of elements a, b of semisimple MV-algebras A and B cannot be defined in terms of the MV-algebraic operations $\neg, \oplus$.

Consistency preserving operations. Generalizing Definition 4.2 *to MV-algebras.* For A and B the independent semisimple MV-subalgebras of the tensor product $A \otimes B$, let α and β be *consistent* $[0, 1]$-valued maps, respectively defined on finite sets $\mathsf{dom}(\alpha) \subseteq A$ and $\mathsf{dom}(\beta) \subseteq B$. With no essential loss of generality the constant function $1 \otimes 1 = 1 \in A \otimes B \supseteq A \cup B$ is a member of $\mathsf{dom}(\alpha) \cap \mathsf{dom}(\beta)$. For any MV-algebra C containing $A \otimes B$ as a subalgebra, the quintuple (A, B, C, α, β) is said to be *amenable*.

A map $\diamond\colon\ [0, 1]^2 \to [0, 1]$ is (A, B, C, α, β)*-consistency preserving* if for all $a, a' \in \mathsf{dom}(\alpha)$, $b, b' \in \mathsf{dom}(\beta)$,

$$a \otimes b = a' \otimes b' \Rightarrow \alpha(a) \diamond \beta(b) = \alpha(a') \diamond \beta(b') \tag{7.15}$$

and the resulting *function* $\alpha \diamond \beta\colon\ a \otimes b \in \mathsf{dom}(\alpha) \otimes \mathsf{dom}(\beta) \mapsto \alpha(a) \diamond \beta(b) \in [0, 1]$ is a consistent joint extension of α and β. Here $\mathsf{dom}(\alpha) \otimes \mathsf{dom}(\beta)$ is shorthand for $\{a \otimes b \in A \otimes B \mid a \in \mathsf{dom}(\alpha),\ b \in \mathsf{dom}(\beta)\}$.

We say that $\diamond$ is *consistency preserving* if it is so for any amenable quintuple (A, B, C, α, β). When this is the case, the map $\alpha \diamond \beta$ is called the $\diamond$*-book of* α *and* β in $A \otimes B$.

The following result extends Theorem 4.3 to all semisimple MV-algebras:

Theorem 7.18 (MV-algebraic product law) *Product is the only continuous* consistency preserving *operation* $\diamond\colon [0,1]^2 \to [0,1]$. *In other words, for all independent semisimple MV-algebras* $A, B \subseteq A \otimes B$ *and consistent maps* α, β *on finite subsets* $\mathsf{dom}(\alpha) \subseteq A$ *and* $\mathsf{dom}(\beta) \subseteq B$ *with* $1 \in \mathsf{dom}(\alpha) \cap \mathsf{dom}(\beta)$ *we have:*

$$\textit{For all } a, a' \in \mathsf{dom}(\alpha),\ b, b' \in \mathsf{dom}(\beta),$$
$$a \otimes b = a' \otimes b' \Rightarrow \alpha(a) \cdot \beta(b) = \alpha(a') \cdot \beta(b'),$$

$\alpha \cdot \beta$ *is a consistent map* $\supseteq \alpha \cup \beta$, *and no other continuous map has these properties.*

Proof (*Uniqueness*) As remarked above, in case A, B, C are boolean algebras our present MV-algebraic definitions of amenability, consistency, independence and state agree with the corresponding definitions given in the first part of the book for boolean algebras. Recalling that boolean algebras coincide with semisimple MV-algebras, by Corollary 4.8 the semisimple tensor product of A and B coincides with the free product $A \otimes B = A \amalg B$. In particular, the $\diamond$ operation is consistency preserving in the sense of the above definition if and only if it is consistency preserving in the sense of Definition 4.2. As a consequence, if the continuous operations $\diamond$ and $\diamond'$ preserve the consistency of all (amenable by construction) quintuples $(A, B, A \otimes B, \alpha, \beta)$ considered in the first part of the proof of Theorem 4.3, the proof shows that $\diamond$ and $\diamond'$ coincide with product over all pairs of rational numbers in $[0, 1]$. Since $\diamond$ is continuous, the proof of uniqueness of $\diamond$ as a continuous consistency preserving operation for this class of quintuples is sufficient to conclude that there is at most one continuous consistency preserving operation in the general sense of the above definition.

(*Existence*) In the light of Theorem 7.4 we may identify A and B with separating MV-algebras of $[0, 1]$-valued continuous functions on the maximal spectral spaces $X \cong \boldsymbol{\mu}(A)$ and $Y \cong \boldsymbol{\mu}(B)$. In symbols,

$$A \subseteq C(X, [0,1]), \quad B \subseteq C(Y, [0,1]). \tag{7.16}$$

Cylindrification induces an isomorphic embedding of A (resp., of B) onto the subalgebra $A_{\mathrm{cyl}\uparrow}$ (resp., onto the subalgebra $B_{\mathrm{cyl}_{\rightarrow}}$) of $A \otimes B$. We may argue identifying A and B with the subalgebras of $C(X \times Y,\ [0, 1])$ given by their cylindrifications. Summing up, A and B are independent subalgebras of $A \otimes B$, and we may write

$$A \otimes B \subseteq C(X \times Y,\ [0,1]) \supseteq A \cup B. \tag{7.17}$$

By (7.16)–(7.17), $A \otimes B$ is a separating MV-subalgebra of $C(X \times Y, [0, 1])$. Furthermore, the compact Hausdorff spaces $X, Y, X \times Y$ are respectively homeomorphic to maximal spectral spaces of A, B, and $A \times B$,

$$\boldsymbol{\mu}(A \otimes B) \cong \boldsymbol{\mu}(A) \times \boldsymbol{\mu}(B) \cong X \times Y.$$

By Theorem 6.3(i)⇒(ii), the consistent maps α, β can be extended to states $\bar{\alpha} \supseteq \alpha$ of A and $\bar{\beta} \supseteq \beta$ of B. Corollary 7.14 then yields two uniquely determined regular Borel probability measures $\bar{\bar{\alpha}} = \kappa(\bar{\alpha})$ and $\bar{\bar{\beta}} = \kappa(\bar{\beta})$ on the compact Hausdorff spaces X and Y satisfying the identities

$$\bar{\alpha}(a) = \int_X a \; \mathrm{d}\bar{\bar{\alpha}} \quad \text{and} \quad \bar{\beta}(b) = \int_Y b \; \mathrm{d}\bar{\bar{\beta}} \quad \text{for all } a \in A \text{and } b \in B.$$

Trivially, both $\bar{\bar{\alpha}}$ and $\bar{\bar{\beta}}$ are σ-finite. Let $\bar{\bar{\mu}}$ be the product measure of the probability measures $\bar{\bar{\alpha}}$ and $\bar{\bar{\beta}}$, as given by Fubini's theorem.[3] Then for each pure tensor $a \otimes b$ of $A \otimes B$ we have

$$\int_{X\times Y} a \otimes b \; \mathrm{d}\bar{\bar{\mu}} = \int_{X\times Y} a_{\mathrm{cyl}\uparrow} \cdot b_{\mathrm{cyl}_{\rightarrow}} \; \mathrm{d}\bar{\bar{\mu}} =$$
$$= \int_Y \left(b \int_X a \; \mathrm{d}\bar{\bar{\alpha}} \right) \mathrm{d}\bar{\bar{\beta}} = \int_X a \; \mathrm{d}\bar{\bar{\alpha}} \cdot \int_Y b \; \mathrm{d}\bar{\bar{\beta}}.$$

Let the map $\bar{\mu} \colon A \otimes B \to [0, 1]$ assign to every element $f \in A \otimes B$ the value $\bar{\mu}(f) = \int_{X\times Y} f \; \mathrm{d}\bar{\bar{\mu}}$. Then $\bar{\mu}$ is a state of $A \otimes B$. Furthermore, for any $a, a' \in A$ and $b, b' \in B$,

$$a \otimes b = a' \otimes b' \Rightarrow \bar{\mu}(a \otimes b) = \bar{\alpha}(a) \cdot \bar{\beta}(b) = \bar{\alpha}(a') \cdot \bar{\beta}(b') = \bar{\mu}(a' \otimes b'). \tag{7.18}$$

Let μ be the restriction of $\bar{\mu}$ to the finite set

$$\{f \in A \otimes B \mid f = a \otimes b = a_{\mathrm{cyl}} \cdot b_{\mathrm{cyl}}, \;\; a \in \mathsf{dom}(\alpha), \;\; b \in \mathsf{dom}(\beta)\}.$$

By (7.18), μ is the product book of α and β,

$$\mu = \alpha \cdot \beta.$$

By Theorem 6.3(ii)⇒(i), μ is consistent. There remains to note that product is continuous and $\mu \supseteq \alpha \cup \beta$, because $1 \in \mathsf{dom}(\alpha) \cap \mathsf{dom}(\beta)$. □

[3] For a proof of Fubini's theorem see the bibliographical notes at the end of this chapter.

In conclusion:

Product is the only continuous $[0, 1]$-valued operation on $[0, 1]^2$ which preserves the consistency of the probabilities of conjunctions of continuous $[0, 1]$-valued events in any two logically independent consistent sets.

7.5 Kolmogorov's Triples and MV-Algebras Enriched with a State

Can Kolmogorov's (expedient and arbitrary by his own admission) countable additivity axiom be reconciled with finitely additive probability?

A first partial positive answer is given by Theorem 4.10, according to which the states of any boolean algebra A correspond, via an affine isomorphism, to the regular Borel probability measures on the Stone space X of A. By Stone duality 4.6, X turns out to be the most general totally disconnected compact Hausdorff space. Then Theorem 4.14 shows that enriched boolean algebras correspond to Kolmogorov triples $(\Omega, \mathcal{B}(\Omega), P)$ *for all boolean spaces* Ω. Still, as already remarked, boolean spaces are a too restricted class. In Theorem 7.20 below we will extend Theorem 4.14, showing that MV-algebras enriched by a state (= a finitely consistent map, by Corollary 6.14) yield all Kolmogorov triples $(\Omega, \mathcal{B}(\Omega), P)$ *for any compact Hausdorff space* Ω. For the proof we prepare:

Definition 7.19 By an *enriched MV-algebra* we mean a pair (A, σ) where A is a semisimple MV-algebra and σ is a state of A. Two enriched MV-algebras (A, σ) and (A', σ') are *isomorphic* if there is an isomorphism $\phi\colon A \to A'$ such that $\sigma = \sigma'(\phi)$, in symbols, $\phi\colon (A, \sigma) \cong (A', \sigma')$.

The affine homeomorphism $\rho\colon \mathsf{S}(A) \cong \mathsf{S}(A/\mathsf{Rad}(A))$ of Proposition 7.5 ensures that we lose nothing of interest in what follows by restricting to semisimple MV-algebras.

For any topological spaces Ω, Ω' and homeomorphism $\psi\colon \Omega \cong \Omega'$ the isomorphism of two Kolmogorov triples $(\Omega, \mathcal{B}(\Omega), P)$ and $(\Omega', \mathcal{B}(\Omega'), P')$ is defined in a similar way.

As above, $C(\Omega, [0, 1])$ denotes the MV-algebra of all continuous $[0, 1]$-valued functions on Ω. Furthermore, for every MV-algebra A,

$$\kappa = \kappa_A\colon \mathsf{S}(A) \cong \mathcal{P}(\boldsymbol{\mu}(A))$$

is the affine isomorphism of Corollary 7.14. We also write $(\Omega, \mathcal{B}(\Omega), P)$ for the Kolmogorov triple, with Ω a *compact Hausdorff* space and $P \in \mathcal{P}(\Omega)$ a regular Borel probability measure on Ω. Throughout, (A, σ) and (A', σ') are enriched

MV-algebras, identified with separating MV-algebras of continuous [0, 1]-valued functions on their own maximal spectral spaces $\boldsymbol{\mu}(A)$ and $\boldsymbol{\mu}(A')$.

Theorem 4.14 has the following generalization:

Theorem 7.20 (Enriched MV-algebras and Kolmogorov triples.)

(i) Let the map $\boldsymbol{\phi}$ *transform the enriched MV-algebra* (A, σ) *into the Kolmogorov triple* $(\boldsymbol{\mu}(A), \mathcal{B}(\boldsymbol{\mu}(A)), \kappa_A(\sigma))$. *Then* $\boldsymbol{\phi}(A, \sigma) \cong \boldsymbol{\phi}(A', \sigma')$ *if and only if there is a homeomorphism* $\iota\colon\ \boldsymbol{\mu}(A) \cong \boldsymbol{\mu}(A')$ *such that for all* $B \in \mathcal{B}(\boldsymbol{\mu}(A))$, $(\kappa_A(\sigma))(B) = (\kappa_{A'}(\sigma'))(\iota(B))$.

(ii) Conversely, let the map $\boldsymbol{\psi}$ *transform any Kolmogorov triple* $(\Omega, \mathcal{B}(\Omega), P)$ *(with* Ω *a compact Hausdorff space) into the enriched MV-algebra given by*

$$(C(\Omega, [0, 1]), \kappa^{-1}_{C(\Omega,[0,1])}(P)).$$

Then $(\Omega, \mathcal{B}(\Omega), P) \cong \boldsymbol{\phi}(\boldsymbol{\psi}(\Omega, \mathcal{B}(\Omega), P))$.

(iii) For any compact Hausdorff space X *and state* σ *of* $C(X, [0, 1])$ *the enriched MV-algebras* $(C(X, [0, 1]), \sigma)$ *and* $\boldsymbol{\psi}(\boldsymbol{\phi}(C(X, [0, 1]), \sigma))$ *are isomorphic. Thus, up to isomorphism, enriched MV-algebras of the form* $(C(X, [0, 1]), \sigma)$ *are in one-one correspondence with Kolmogorov triples* $(\Omega, \mathcal{B}(\Omega), P)$, *for any compact Hausdorff space* Ω.

Proof From Proposition 7.1, Theorem 7.4, and Corollaries 7.12 and 7.13. □

Remarks 7.21 While de Finetti's consistency notion is generally considered a cornerstone of finitely additive probability, Theorem 7.20 highlights the deep relations between finitely consistent maps (i.e., states) σ on MV-algebras A, and Kolmogorov triples $(\Omega, \mathcal{B}(\Omega), P) \cong (\boldsymbol{\mu}(A), \mathcal{B}(\boldsymbol{\mu}(A)), \kappa(\sigma)) = \boldsymbol{\phi}(A, \sigma)$ on any compact Hausdorff space Ω. The theorem shows that Kolmogorov's "expedient" and "arbitrary" countable additivity axiom is the counterpart of a fundamental consistency notion, which is in turn equivalent to extendability to a state (= finitely consistent map). In a sense, Kolmogorov triples $(\boldsymbol{\mu}(A), \mathcal{B}(\boldsymbol{\mu}(A)), \kappa(\sigma))$ stand to MV-algebras $C(\boldsymbol{\mu}(A), [0, 1])$ enriched by a state σ as integration theory on $\mathcal{B}(\boldsymbol{\mu}(A))$ stands to normalized positive continuous linear functionals (Radon measures) on $C(\boldsymbol{\mu}(A), \mathbb{R})$.

Once continuous [0, 1]-valued random variables are *coded* by formulas in Łukasiewicz logic they can be acted upon by the natural algorithmic apparatus of this logic. One can then construct *algorithms and decision procedures*, to check the consistency, the logic independence, and other basic properties of the probabilities assigned to these random variables in many real-life situations, not necessarily restricted to bookmaking. This is the subject matter of the next two chapters.

To the attention of readers interested in further developments of the MV-algebraic approach to Kolmogorov probability it may be appropriate to recall here Rota's introduction to his 1998 Fubini lectures:

> The purpose of this lecture, and of the lectures that will follow it, is a controversial one. It is, plainly and simply, to reopen some questions in the foundations of probability, and thus also a much touchier and much neglected subject, the foundations of statistics. There seems to have been a conspiracy of silence on such foundational questions among mathematicians (not so among philosophers), who would like to believe that the subject has been definitively closed by Kolmogorov, and many of whom react unfavorably to any suggestion of revision of Kolmogorov's elegant setup.
>
> [...]
>
> I will lay my cards on the table: a revision of the notion of a sample space is my ultimate concern. I hasten to add that I am not about to put forth concrete proposals for carrying out such a revision. We will, however, be guided by a belief that has been a guiding principle of the mathematics of this century. Analysis will play second fiddle to algebra. The algebraic structure sooner or later comes to dominate, whether or not it is recognized when a subject is born. Algebra dictates the analysis.

Historical and bibliographical notes. Rota's opening remarks in his Fubini Lectures can be found in [18, p. 57].

Vector lattices are extensively described in [22]. In particular, see [22, p. 372] for unital vector lattices. Details on divisible hulls of ordered groups can be found in [2, 1.6.8–1.6.9]. For a proof of the lattice version of the Stone-Weierstrass theorem see [10, Theorem 7.29] or [22, pp. 8–10]. For the Riesz-Markov-Kakutani representation theorem see [19, 2.14, 6.19] or [22, p. 119 and references therein]. For Fubini's theorem see [1, p. 233].

Theorem 7.13 was first proved in [12, Corollary 29]. The straightforward generalization to all MV-algebras immediately follows from the affine homeomorphism of Proposition 7.5(ii), see [17, bottom lines of page 2850].

With reference to Theorem 7.20(iii), MV-algebras of the form $C(X, [0, 1])$ for some compact Hausdorff space X can be characterized as the Γ-correspondents of archimedean *divisible norm-complete* unital ℓ-groups. This follows from [9, Theorem 5.5] using the preservation properties of Γ. It may be noted that MV-algebras $C(X, [0, 1])$ naturally arise from the adjunction constructed in [5, 4.7] between MV-algebras and the C*-algebraic dual of the category of compact Hausdorff spaces.

For an example showing that the free product of two MV-algebras differs from their tensor product see [15, Example 9.15, p.108]. In the same paper the reader will also find the standard definition of MV-algebraic tensor product in terms of universal bimorphisms. See [15, Theorem 9.17] for the canonical isomorphism of the semisimple tensor product $A \otimes B$ onto the MV-algebra of $[0, 1]$-valued functions on the cartesian product $X \times Y$ generated by the functions of the form $f \cdot g$, for all $f \in A \subseteq C(X, [0, 1])$ and $g \in B \subseteq C(Y, [0, 1])$. This yields the handy definition of semisimple tensor product adopted in this chapter.

For different approaches to algebraic probability in the literature see, e.g., [21] and [20]. In [16, §§2 and 4] Nelson advocates a radically elementary ring-theoretic approach to probability. By Gelfand duality, [11], the ring-theoretic nature of MV-algebras follows from the above mentioned adjunction constructed in [5, 4.7]. From a different viewpoint, let $\mathfrak{A}$ be a Bratteli AF-algebra, [4], i.e., the norm-closure of the union of a finite ascending chain of finite-dimensional C*-algebras, all with the same unit. Let us suppose that the Grothendieck group $(K_0(\mathfrak{A}), K_0(\mathfrak{A})^+, [1_{\mathfrak{A}}])$ is lattice-ordered, for short, $\mathfrak{A}$ is an *AFℓ-algebra*. Then $\Gamma(K_0(\mathfrak{A}))$ is the most general possible countable MV-algebra. Commutative AF-algebras are automatically AFℓ-algebras and correspond to countable boolean algebras. Interestingly enough, the composite functor $\Gamma(K_0)$ yields a bijection between AFℓ-algebras with a distinguished tracial state and countable enriched MV-algebras. This follows combining Elliott classification [6], [7] for the special case of AFℓ-algebras, [13, 14] with the affine homeomorphism in [3, 3.2] (also see [8, p. 310]).

References

1. Billingsley, P. (1995). Probability and measure (3rd ed.). Wiley.
2. Bigard, A., Keimel, K., & Wolfenstein, S. (1977). Groupes et anneaux réticulés. Lecture Notes in Mathematics (Vol. 608). Springer.
3. Blackadar, B. E. (1980). Traces on simple AF C*-algebras. *Journal of Functional Analysis, 38*, 156–168.
4. Bratteli, O. (1972). Inductive limits of finite dimensional C*-algebras. *Transactions of the American Mathematical Society, 171*, 195–234.
5. Cignoli, R. L. O., Dubuc, E. J., & Mundici, D. (2004). Extending Stone duality to multisets and locally finite MV-algebras. *Journal of Pure and Applied Algebra, 189*, 37–59.
6. Elliott, G. A. (1976). On the classification of inductive limits of sequences of semisimple finite-dimensional algebras. *Journal of Algebra, 38*(1), 29–44.
7. Goodearl, K. R. (1982). Notes on real and complex C*-algebras, Shiva mathematics series (Vol. 5). Shiva Publishing Ltd.
8. Goodearl, K. R. (1986). *Partially ordered Abelian groups with interpolation*. American Mathematical Society.
9. Goodearl, K. R., & Handelman, D. E. (1980). Metric completions of partially ordered abelian groups. *Indiana University Mathematical Journal, 29*, 861–895.
10. Hewitt, E., & Stromberg, K. (1965). *Real and abstract analysis*. Springer.
11. Johnstone, P. (1982). Stone spaces, Cambridge studies in advanced mathematics (Vol. 3). Cambridge University Press. Reprinted (1986).
12. Kroupa, T. (2006). Every state on a semisimple MV-algebra is integral. *Fuzzy Sets and Systems, 157*, 2771–2782.
13. Mundici, D. (1986). Interpretation of AF C^*-algebras in Łukasiewicz sentential calculus. *J. Functional Analysis, 65*, 15–63.
14. Mundici, D. (2023). AF-algebras with lattice-ordered K_0: Logic and computation. Annals of Pure and Applied Logic, 174(1), Article 103182.
15. Mundici, D. (2011). Advanced Łukasiewicz calculus and MV-algebras, Trends in Logic (Vol. 35). Springer.
16. Nelson, E. (1987). Radically elementary probability theory, Annals of Mathematics Studies (Vol. 117).
17. Panti, G. (2009). Invariant measures in free MV-algebras. *Communications in Algebra, 36*, 2849–2861.
18. Rota, G.-C. (2001). Twelve problems in probability no one likes to bring up. The Fubini Lectures, Torino, 3–5 June, 1998. In H. Crapo & D. Senato (Eds.), Algebraic combinatorics and computer science, a tribute to Gian-Carlo Rota (pp. 57–93). Springer, Milan. https://link.springer.com/chapter/10.1007/978-88-470-2107-5_5.
19. Rudin, W. (1987). Real and complex analysis (3rd ed.). McGraw-Hill Book Co.
20. Rusza, I. Z., & Szekely, G. J. (1988). *Algebraic probability theory*. Wiley.
21. Tao, T., Algebraic probability spaces. https://terrytao.wordpress.com/2014/06/28/algebraic-probability-spaces/.
22. Yosida, K. (1980). Functional analysis (6th ed.). Springer.

Chapter 8
Deciding PSAT_∞ for [0, 1]-Valued Continuous Events

Every day we bet on weather conditions, waiting times, the price of oil, etc. As we have seen, MV-algebras stand to these events as boolean algebras stand to the yes-no events considered in the first part of the book. That's why Bookie has expanded his business, and now also sells/buys bets on continuous [0, 1]-events. Over the years, Bookie's ideal booking has become increasingly popular, to the point that today there is even an Ideal Booking Society. All its members are equipped with a gadget checking the consistency and inconsistency of books. This chapter deals with the algorithms running in these devices.

The reader is supposed to have some familiarity with the rudiments of computation theory. Whatever we say here about Turing machines, decidability, algorithms ..., we shall assume it outright and ask the interested reader to look up the matter in the references mentioned at the end of this chapter. However, there should be no question about what we mean when we say that some procedure/proof/problem is effective/constructive/decidable.

8.1 Examples

At the end of Chapter 7 it was proved that MV-algebras and their states yield all Kolmogorov triples $(\Omega, \mathcal{B}(\Omega), P)$ on any compact Hausdorff space Ω. One may now wonder what Łukasiewicz logic $Ł_\infty$, MV-algebras and their states have to offer to probability theory. In this chapter we devise an *algorithm* that decides the PSAT_∞ problem, a generalization for continuous events of the PSAT problem. An input instance of PSAT_∞ is a finite set E of k-variable $Ł_\infty$-formulas, where each formula $\psi_j \in E$, $(j = 1, \ldots, m)$ is assigned a rational number $r_j \in [0, 1]$. Intuitively, r_j is the "betting odd" assigned by Bookie to ψ_j. Any formula ψ_j in Bookie's book codes a McNaughton function $\mathfrak{f}_j \in \mathcal{M}([0, 1]^k)$, representing a continuous event. The PSAT_∞ problem asks whether the set of pairs $\{(\mathfrak{f}_1, r_1), \ldots, (\mathfrak{f}_m, r_m)\}$ is *consistent*,

D. Mundici, *Algebraic Probabilistic Consistency*, Trends in Logic 69,
https://doi.org/10.1007/978-3-031-98337-5_8

or equivalently, whether this set can be extended to a state of the free MV-algebra $\mathcal{M}([0,1]^k)$. Our main result in this chapter states that the PSAT$_\infty$ problem is Turing-decidable.

A first example. Let us assume Bookie's book assigns betting odds to the possible results of the forthcoming match between the national football teams W and V. Bookie only accepts bets on the timing of the team's first goal under the assumption that team W will win. (If W does not win, the cash bet will be returned.) Here are some examples from Bookie's book:

"W's first goal will be scored at the outset of the game", 0.34

"W's first goal will be scored towards the end of the match", 0.16

"W's first goal will be scored at the end of first half", 0.53.

For instance, Bookie precisely defines the first event in the following diagram:

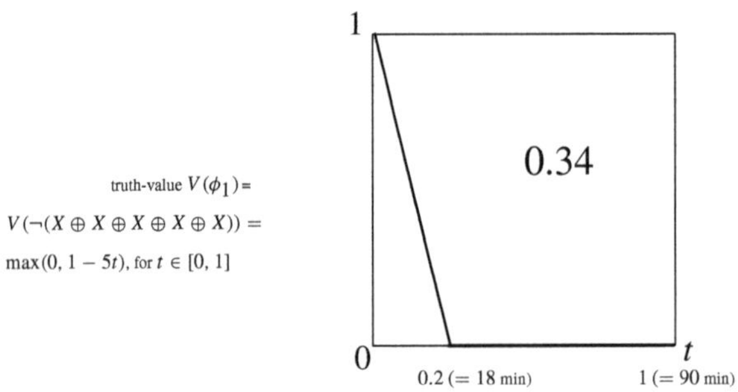

Bookie's definition of the event "team W will score at the outset of the game", by means of the Łukasiewicz formula $\neg(X \oplus X \oplus X \oplus X \oplus X)$ and the graph of its McNaughton function. Bookie's betting odd for this event is equal to 0.34.

Similar diagrams are posted for all other events in Bookie's book. The interval $[0, 1]$ in abscissa stands for the normalized length of the match. The rational numbers 0.34, 0.16, 0.53,...are Bookie's betting odds for these events. Each event ψ comes with a one-variable Ł$_\infty$-formula ψ assigning the truth-value $\mathsf{f}_\psi(t)$ to any possible occurrence $t \in [0, 1]$ of the first goal scored by team W.

Thus in particular, for the event $\psi_1 =$ "W will score an early goal" the McNaughton function $\mathsf{f}_{\psi_1}\colon [0,1] \to [0,1]$ coded by the formula $\psi_1 = \neg(X \oplus X \oplus X \oplus X \oplus X)$ takes value 1 at $t = 0$. The event ψ_1 is 100% true if W scores a goal in the first seconds of the game. Then the truth-value function $\mathsf{f}_{\psi_1}(t)$ descends linearly to zero until $t = 1/5$ (= the 18th minute). For all $t \in [1/5, 1]$ the truth-value $\mathsf{f}_{\psi_1}(t)$ is constantly equal to 0. Summing up, $\mathsf{f}_{\psi_1}(t) = \max(0, 1 - 5t)$ for any $t \in [0, 1]$. The $Ł_\infty$-formula ψ_1 is Bookie's precise definition of the first event in his book.

The pairs (ψ_j, r_j), $j = 1, \ldots, m$, *precisely determine the betting rules.* Thus for instance, if Betty bets s_1 (euro) on ψ_1, now she pays Bookie $s_1 \cdot 0.34 = s_1 \cdot r_1$. She will pay back $s_1 \cdot \mathsf{f}_{\psi_1}(\bar{t})$ when the timing $\bar{t}$ of W's first goal is known. In the best tradition of Bookie's ideal business, Betty is allowed to swap roles with Bookie, by just making a *negative bet* $s_1 < 0$ on ψ_1. Then Bookie pays Betty $|s_1| \cdot 0.34$, and will be paid back $|s_1| \cdot \mathsf{f}_{\psi_1}(\bar{t})$ when the value $\bar{t}$ is known. With the usual bettor-to-bookie orientation, for any $s_1 \in \mathbb{R}$ the resulting financial transaction is given by $s_1 \cdot (r_1 - \mathsf{f}_{\psi_1}(\bar{t}))$.

> The problem is whether Bookie's book $\{(\psi_1, r_1), \ldots, (\psi_m, r_m)\}$ is inconsistent. In other words, can Betty make a bet $s_1, \ldots, s_m$ that will ruin Bookie regardless of the timing of W's first goal? More precisely, do there exist $s_1, \ldots, s_m \in \mathbb{R}$ satisfying the inequality $\sum_{j=1}^{m} s_j \cdot (r_j - \mathsf{f}_{\psi_j}(\bar{t})) < 0$ for every $\bar{t} \in [0, 1]$?

The reader may wonder why only McNaughton functions are chosen to deal with [0, 1]-events, rather than, say, polynomials or other classes of continuous functions. A main reason is that Łukasiewicz logic comes with a natural logic-algorithmic machinery allowing us to *mechanically* recognize inconsistent books of betting odds. Moreover, McNaughton functions restricted to arbitrary finite unions of rational simplexes in $[0, 1]^n$ have a sufficient *expressive power* for the most general conditional PSAT_∞ problem. This will become apparent in the next chapter.

A second example. Let us consider a book whose 5 events deal with the ratio $q = \frac{\textit{gold price today}}{\textit{thirty days from now}}$ between the price of gold today at 12 GMT, and its price thirty days from now, at 12 GMT. Bookie's book only sells/buys bets under the hypothesis that the price of gold will increase, i.e., $q \in [0, 1]$. As in the previous example, if the ratio decreases, the cash bet will be returned. Bettors are faced with the following list of *imprecisely* defined options and corresponding betting odds:

1. q will be more or less stable: $r_1 = 0.70$
2. q will not dramatically decrease below 0.75: $r_2 = 0.82$
3. q will be around 0.75: $r_3 = 0.15$
4. q will be around 0.60: $r_4 = 0.13$
5. q will dramatically drop below 0.5: $r_5 = 0.09$.

For each $j = 1, \ldots, 5$ the description of the jth event in the above list comes with a rational betting odd $r_j \in [0, 1]$ and a one-variable $Ł_\infty$-formula ψ_j coding a McNaughton function $\mathsf{f}_j\colon [0, 1] \to [0, 1]$ whose graph is also illustrated in the book as a visual aid to the bettor:

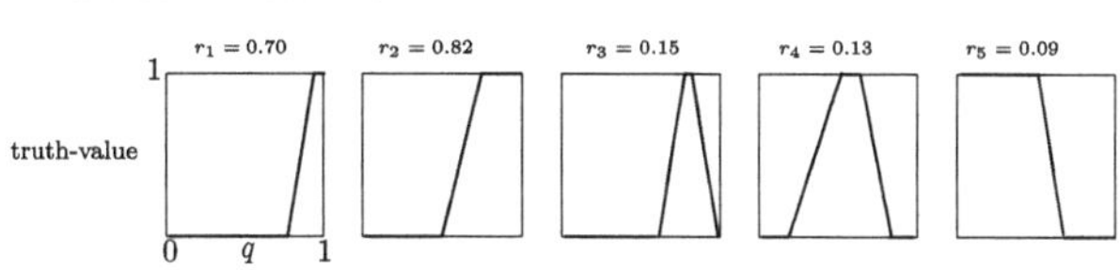

As in the previous example, for any $q \in [0, 1]$ the real number $\mathsf{f}_j(q) \in [0, 1]$ is the truth-value of the event coded by ψ_j if the ratio of gold prices is q thirty days from now. It is understood that the betting rules are determined solely by the pairs $(\psi_j, r_j),\ j = 1, \ldots, 5$.

Again, the problem is whether the set of pairs $\beta = \{(\mathsf{f}_1, r_1), \ldots, (\mathsf{f}_5, r_5)\}$ is consistent. In Definition 8.1 below we will define the consistency of β in the usual way. By Theorem 6.3 we already know that the consistency of β is equivalent to its extendability to a state of the free MV-algebra $\mathcal{M}([0, 1]^k)$.

More generally, Bookie's books can also deal with a finite set of k-dimensional events, for some fixed $k = 1, 2, \ldots$. In this chapter we are only concerned with books on events coded by formulas in k-variables, whose McNaughton functions are defined over the unit k-cube $[0, 1]^k$. In the next chapter we will consider more general situations where the events are constrained by a finite set of conditions, also defined by formulas in Łukasiewicz logic.

8.2 Statement of the PSAT$_\infty$ Problem

A preliminary condition to the decidability of the consistency of Bookie's book is that this can be presented as a finite string of input symbols for a Turing machine. As in our examples above, this input will consist of a finite list of pairs $(\psi_j, r_j),\ j = 1, \ldots, m$, with each ψ_j a k-variable Ł$_\infty$-formula, and r_j a rational number in the unit interval, typically coded as a pair of integers $a_j \in \mathbb{Z}_{\geq 0}$ and $b_j \in \mathbb{Z}_{>0}$ in binary notation, with $b_j \geq a_j$. We let f_j be the McNaughton function coded by ψ_j.

Definition 8.1 The *PSAT$_\infty$ decision problem* is as follows:

INPUT k-variable Ł$_\infty$-formulas $\psi_1, \ldots, \psi_m$ and rationals $r_1, \ldots, r_m \in [0, 1]$.

QUESTION Is the set $\beta = \{(\mathsf{f}_1, r_1), \ldots, (\mathsf{f}_m, r_m)\}$ consistent? In other words, does every $(s_1, \ldots, s_m) \in \mathbb{R}^m$ admit an $\eta \in \mathsf{hom}(\mathcal{M}([0, 1]^k))$ such that $\sum_{j=1}^m s_j(r_j - \eta(\mathsf{f}_j)) \geq 0$? Equivalently (by Theorem 6.3), is β extendable to a state of $\mathcal{M}([0, 1]^k)$?

Our main result in this chapter is as follows:

Theorem 8.2 *The PSAT$_\infty$ problem is Turing-decidable.*

The proof of this theorem will take the remainder of the chapter.

Remark 8.3 We must preliminarily check the following conditions ensuring the grammatical correctness of the input string $\{(\psi_1, r_1), \ldots, (\psi_m, r_m)\}$:

1. Each string r_j must code a rational in the unit real interval.
2. Each string ψ_j must be a grammatically correct Ł$_\infty$-formula.

For (1), on any input string consisting of two natural numbers written in binary notation it is easy to check whether the string denotes a rational in [0, 1]. Likewise, for (2), a parsing algorithm quickly checks if an input formula ψ_j is correctly written. ($Ł_\infty$-formulas have the same inductive definition as boolean formulas). If any of these checks fails, the decision procedure is called off. If both checks are positive, to avoid Bookie's premature bankrupt we must now check that logically equivalent formulas in the input list $\psi_1, \ldots, \psi_m$ must be assigned equal betting odds. More precisely our input must satisfy the following condition:

(3) Whenever $\mathsf{f}_u = \mathsf{f}_v$, the betting odds r_u and r_v must coincide. For otherwise, Betty bets a zillion on the cheaper input formula ψ_u and sells the more expensive ψ_v to Bookie for a zillion. Whatever happens, Bookie will go bankrupt. That's why to decide PSAT_∞ we must preliminarily check that the set of pairs $\beta = \{(\mathsf{f}_1, r_1), \ldots, (\mathsf{f}_m, r_m)\}$ is a *function*. Recall Remark 3.15.

The following nontrivial proposition ensures that also this preliminary check can be carried out effectively:

Proposition 8.4 *There is a Turing machine $T_{\text{logically-equivalent}}$ to recognize $Ł_\infty$-equivalent formulas. Thus a trivial refinement of $T_{\text{logically-equivalent}}$ checks that any two equivalent formulas ψ_j, ψ_l are assigned equal betting odds.*

Proof This requires an adaptation of the co-NP-complete recognition algorithm for tautologies of the form $\phi \leftrightarrow \psi$. A Turing machine performing this task exists by Theorem 4.6.10 in TheBook. The rest immediately follows. □

In case two equivalent formulas are assigned different betting odds the decision procedure outputs "inconsistent input". If any two equivalent $\psi_j, \psi_{j'}$ formulas are assigned equal odds $r_j, r_{j'}$, we let a refinement of $T_{\text{logically-equivalent}}$ delete the pair $(\psi_{j'}, r_{j'})$ and proceed until all input formulas are pairwise inequivalent.

Having thus ensured that the resulting set of pairs $\{(\mathsf{f}_1, r_1), \ldots, (\mathsf{f}_m, r_m)\}$ is a *function*, it makes sense to ask about its extendability to a state.

However, a further preliminary check is advisable, if only to warn Bookie of an impending doom:

(4) For any $j = 1, \ldots, m$ we must have $r_j \leq \max\{\mathsf{f}_j(x) \mid x \in [0, 1]^k\}$. For otherwise, Betty sells this event to Bookie for a zillion and, whatever the outcome of the event, Bookie will go bankrupt. Symmetrically, we must check that $r_j \geq$ the minimum value of f_j over $[0, 1]^k$.

The decidability of the inequalities in (4) is a nontrivial result, to be proved in Lemma 8.8. To proceed, we next prepare the necessary material in rational polyhedral topology.

8.3 Constructions on Rational Polyhedral Sets

A point $y \in \mathbb{R}^n$ is *rational* if all its coordinates are rational. For convenience, in this chapter we will use the notation

$$a \circ b \quad \text{rather than} \quad a^T b \quad \text{used in earlier sections} \tag{8.1}$$

for the scalar product of vectors $a, b \in \mathbb{R}^n$. By a *rational hyperplane* H in $\mathbb{R}^n$ we mean a set of the form $H = \{x \in \mathbb{R}^n \mid h \circ x = c\}$, for some nonzero vector $h \in \mathbb{Q}^n$ and constant $c \in \mathbb{Q}$. We say that h is a *normal vector of* H. When $c = 0$ we say that H is *a homogeneous linear hyperplane*. The two closed halfspaces H^+ and H^- of $\mathbb{R}^n$ with boundary H are said to be *rational.*

Recall that for $0 \leq m \leq n$, an *m-simplex* in $\mathbb{R}^n$ is the convex hull $T = \mathsf{conv}(w_0, \ldots, w_m)$ of $m + 1$ affinely independent points in n-dimensional euclidean space $\mathbb{R}^n$. The *vertices* $w_0, \ldots, w_m$ of T are uniquely determined by T. T is *rational* if so is each vertex w_i. The empty set is the only -1-simplex.

A (rational) *polytope* P in $\mathbb{R}^n$ is the convex hull of a finite set of (rational) points in $\mathbb{R}^n$. Minkowski's n-dimensional version of the Krein-Milman theorem states that P is the convex hull of its vertices. Recalling Lemma 2.3, by a *face* F of P we mean either P itself, or the empty set $\emptyset$, or the intersection of P with a supporting hyperplane of P. P and $\emptyset$ are the *improper faces* of P. By a *facet* we mean an $n-1$-dimensional face. A *polyhedral set* (also known as an "$\mathcal{H}$-polyhedron", or a "k-dimensional polyhedron", or just a "polyhedron") is a subset Q of $\mathbb{R}^k$ which is the intersection of finitely many closed halfspaces in $\mathbb{R}^k$, or $Q = \mathbb{R}^k$ itself.

Theorem 8.5 *If $P \subseteq \mathbb{R}^n$ is a bounded polyhedral set then P is a polytope.*

Proof For halfspaces $K_1, \ldots, K_l$ we may assume that $P = \bigcap_{t=1}^{l} K_t$ is n-dimensional, with $n \geq 2$. Trivially, P is closed and convex. Since P is also bounded, by Minkowski's (Krein-Milman) Theorem, P is the convex hull of the set $\mathsf{ext}\, P$ of its extreme points (also called vertices),

$$P = \mathsf{conv}\,\mathsf{ext}\, P. \tag{8.2}$$

There remains to prove that P has only finitely many extreme points. By hypothesis, P is the intersection of a set

$\mathcal{K} = \{K_1, \ldots, K_l\}$ of halfspaces of $\mathbb{R}^n$.

Let

$\mathcal{I} = \{I_1, \ldots, I_l\}$ be the set of linear inequalities specifying these halfspaces,

$\mathcal{H} = \{H_1, \ldots, H_l\}$ be the set of hyperplanes bounding the halfspaces of $\mathcal{K}$,

$\mathcal{E} = \{E_1, \ldots, E_l\}$ be the set of linear equalities specifying these hyperplanes.

For any $x \in P$ let $rank(x)$ be the maximum number of linearly independent equalities in $\mathcal{E}$ satisfied by x. Thus, e.g., if $rank(x) = 0$, x lies in the interior of P. If $1 \leq$

$rank(x) < n$, x lies on a line L and so does any point y in a suitably small relatively open segment $O \subseteq L$ containing x. It follows that x is not an extreme point in P.

We conclude that any extreme point of P is the unique solution of a system of n linearly independent equations in the finite set $\mathcal{E}$. Different extreme points of P are specified by different subsets of $\mathcal{E}$. At least one extreme point of P exists by (8.2). There are only finitely many elements in $\mathcal{E}$. □

Corollary 8.6 *There exists a Turing machine which, having in input a bounded rational polyhedral set P outputs the extreme points of P.*

Proof With reference to the proof of Theorem 8.5, the coefficients of any equation in the finite set $\mathcal{E}$ are rational. From the constructive character of the proof it follows that all extreme points of P are Turing computable rational numbers. □

Also the converse of Theorem 8.5 holds, but since it will find no use in this book its proof is omitted. The interested readers will be addressed to the literature when many different proofs of the converse, as well as of the above theorem can be found.

A *polyhedral complex* in $\mathbb{R}^n$ (also known as a "polytopal complex", or just a "complex") is a nonempty finite set $\mathcal{C}$ of polytopes in $\mathbb{R}^n$ such that the face of any $P \in \mathcal{C}$ is a member of $\mathcal{C}$, and for any $P, Q \in \mathcal{C}$, the intersection $P \cap Q$ is a common face of P and Q.

We say that $\mathcal{C}$ is *rational* if so are all its members. $\mathcal{C}$ is a *simplicial complex* if all its elements are simplexes.

For any polyhedral complex $\mathcal{C}$ the pointset union of all its elements is said to be the *support* of $\mathcal{C}$, denoted $|\mathcal{C}|$. In case $\mathcal{C}$ is a simplicial complex, $\mathcal{C}$ is said to be a *triangulation* of $|\mathcal{C}|$. A polyhedral complex $\mathcal{C}'$ is a *subdivision of* $\mathcal{C}$ if $|\mathcal{C}'| = |\mathcal{C}|$ and every member of $\mathcal{C}$ is a union of members of $\mathcal{C}'$.

8.4 A Machine that Linearizes the McNaughton Functions $\mathsf{f}_1, \ldots, \mathsf{f}_m$

Lemma 8.7 *There is a Turing machine $T_{\mathsf{linearizer}}$ which, on input consisting of k-variable formulas $\psi_1, \ldots, \psi_m$* linearizes *the function $\mathsf{f}_{\psi_j} \colon [0,1]^k \to [0,1]$ for each $j = 1, \ldots, m$. In other words, $T_{\mathsf{linearizer}}$ outputs a rational triangulation ∇ of $[0,1]^k$ such that each McNaughton function f_{ψ_j} is linear (always in the affine sense) over each simplex of ∇.*

Proof Let f_j be shorthand for f_{ψ_j}.

First Part: Computing the linear pieces of f_j, for each $j = 1, \ldots, m$.

There is a Turing machine $T_{\mathsf{list\ linear\ pieces}}$ which, having in input a formula ψ in the variables $X_1, \ldots, X_k$, outputs a list of linear polynomials with integer coefficients containing all linear pieces of f_ψ. $T_{\mathsf{list\ linear\ pieces}}$ works by induction on the number of occurrences of constant symbols and connectives in ψ as follows:

Basis: The linear polynomials for the linear pieces of (the McNaughton functions associated to) the constant symbol 0 and the variables $X_1, \ldots, X_k$ are the constant function zero and the projections $\pi_1, \ldots, \pi_k$ respectively.

Induction step: If $p_1, \ldots, p_r$ are the linear pieces of ψ, then $1 - p_1, \ldots, 1 - p_r$ are the linear pieces of $\neg\psi$. If $f_1, \ldots, f_u$ are the linear pieces of ϕ and $g_1, \ldots, g_v$ are the linear pieces of γ then the linear pieces of $\phi \oplus \gamma$ are the constant function 1 together with the polynomials $f_l + g_n$ for all $l = 1, \ldots, u$ and $n = 1, \ldots, v$. Applying this procedure to each input formula ψ_j, the machine $T_{\text{list linear pieces}}$ outputs the linear pieces

$$p_{j,1}, \ldots, p_{j,w_j}$$

of the McNaughton function f_j coded by ψ_j, for every $j = 1, \ldots, m$.

Second Part: Linearizing f_j *for every* $j = 1, \ldots, m$. (Compare with the analysis in TheBook, §3.3 and in the proof of Proposition 9.1.4 therein.)

Let $\{p_1, \ldots, p_i\} = \bigcup\{\{p_{j,1}, \ldots, p_{j,w_j}\} \mid j = 1, \ldots, m\}$. For every permutation π of the index set $\{1, \ldots, i\}$ let

$$P_\pi = \{y \in [0, 1]^k \mid p_{\pi(1)}(y) \leq \cdots \leq p_{\pi(i)}(y)\}.$$

By Theorem 8.5, every nonempty P_π is a rational polytope, possibly of dimension $< k$. By Corollary 8.6, some Turing machine on input $\{p_1, \ldots, p_i\,;\,\pi\}$ outputs the extreme points of P_π. Any two distinct polytopes in the set $\mathcal{K}$ of k-dimensional polytopes of the form P_π have disjoint interiors. By direct inspection of the constructive character of the proof of Theorem 8.5, some Turing machine $T_{\text{triangulate}}$ outputs a rational triangulation ∇ of $[0, 1]^k$ such that every k-dimensional polytope P_π of $\mathcal{K}$ is a union of k-simplexes of ∇. By adding a maximal set of diagonals, we may insist that no new vertices are added by $T_{\text{triangulate}}$ in this triangulation. By construction, every function f_j is linear over every simplex of ∇. □

As promised, to complete our preliminary check of the input string of PSAT$_\infty$, we can now show the decidability of the two conditions stated in (4) at the end of Sect. 8.2.

Lemma 8.8 *There is a Turing machine to check that for any* $j = 1, \ldots, m$ *the rational* r_j *associated to* ψ_j *satisfies the inequalities* $\min \mathsf{f}_j \leq r_j \leq \max \mathsf{f}_j$ *over the* k*-cube* $[0, 1]^k$.

Proof For any fixed $j = 1, \ldots, m$, it is enough to check the inequalities $\min \mathsf{f}_j \leq r_j \leq \max \mathsf{f}_j$ by restricting the search of $\min \mathsf{f}_j$ and $\max \mathsf{f}_j$ from the set of all $x \in [0, 1]^k$ to the finite set x_t, $(t = 1, \ldots, w)$ of vertices the rational triangulation ∇ of $[0, 1]^k$ constructed by $T_{\text{linearizer}}$. This can be done effectively, because the value $\mathsf{f}_j(x_t)$ is computable from ψ_j for each $t = 1, \ldots, w$ and is effectively comparable with the rational r_j. □

8.5 Conclusion of the Proof of the Decidability of PSAT$_\infty$

After the proof of Lemma 8.8, we may proceed assuming that the input string $\{(\psi_1, r_1), \ldots, (\psi_m, r_m)\}$ has passed the preliminary checks (1)-(4) spelled out in Section 8.2. Otherwise, the input is either grammatically wrong or trivially inconsistent. The Turing machine $T_{\mathsf{linearizer}}$ has just built a rational triangulation ∇ such that every function $\mathsf{f}_1, \ldots, \mathsf{f}_m \in \mathcal{M}([0, 1]^k)$ is linear over each simplex of ∇. Let

$$x_1, \ldots, x_w \text{ be the vertices of } \nabla.$$

For each $x \in [0, 1]^k$ we let the homomorphism $\eta_x \in \mathsf{hom}\,\mathcal{M}([0, 1]^k)$ correspond to x via the homeomorphisms

$$\mathsf{val}\colon x \in [0, 1]^k \mapsto \eta_x \in \mathsf{hom}(\mathcal{M}([0, 1]^k)) \text{ and } \dot{Z}\mathsf{ker}\colon \mathsf{hom}(\mathcal{M}([0, 1]^k)) \to [0, 1]^k$$

of Proposition 7.7. The following lemma provides the key step in the proof of Theorem 8.2.

Lemma 8.9 *Let $s = (s_1, \ldots, s_m) \in \mathbb{R}^m$ satisfy the following condition:*

$$(\nabla\text{-}inconsistency)\ \ For\ each\ t = 1, \ldots, w,\ \ \sum_{j=1}^{m} s_j(r_j - \eta_{x_t}(\mathsf{f}_j)) < 0.$$

Then the input string $\{(\psi_1, r_1), \ldots, (\psi_m, r_m)\}$ is inconsistent, *in the sense that*

$$for\ each\ \eta \in \mathsf{hom}(\mathcal{M}([0, 1]^{\mathsf{k}})),\ \ \sum_{j-1}^{m} s_j(r_j - \eta(\mathsf{f}_j)) < 0.$$

Proof As in (8.1), let $\circ$ denote scalar product in the vector space $\mathbb{R}^m$. ∇-inconsistency amounts to the existence of a solution $s = (s_1 \ldots, s_m)$ of the following system of strict linear inequalities, in correspondence with the homomorphisms $\eta_{x_1}, \ldots, \eta_{x_w}$:

$$\begin{array}{l}(s_1, \ldots, s_m) \circ (r_1 - \eta_{x_1}(\mathsf{f}_1), \ldots, r_m - \eta_{x_1}(\mathsf{f}_m)) < 0\\ (s_1, \ldots, s_m) \circ (r_1 - \eta_{x_2}(\mathsf{f}_1), \ldots, r_m - \eta_{x_2}(\mathsf{f}_m)) < 0\\ \quad\ldots \qquad \ldots \qquad\qquad \ldots \qquad\qquad \ldots\\ (s_1, \ldots, s_m) \circ (r_1 - \eta_{x_w}(\mathsf{f}_1), \ldots, r_m - \eta_{x_w}(\mathsf{f}_m)) < 0.\end{array}$$

Geometrically, this system of inequalities says that for each $t = 1, \ldots, w$ the vector $s \in \mathbb{R}^m$ forms an *obtuse* angle with the vector $d_t = d_{x_t} \in \mathbb{R}^m$ given by

$$d_t = (r_1 - \eta_{x_t}(\mathsf{f}_1), \ldots, r_m - \eta_{x_t}(\mathsf{f}_m)), \tag{8.3}$$

i.e., with the $(r_1, \ldots, r_m)$-translate of the vector $(-\eta_{x_t}(\mathsf{f}_1), \ldots, -\eta_{x_t}(\mathsf{f}_m)) \in \mathbb{R}^m$.

We must show that the above inequalities involving $x_1, \ldots, x_w$ and $\eta_{x_1}, \ldots, \eta_{x_w}$ arising from the assumed ∇-inconsistency still hold over the infinite set of homomorphisms $\eta = \eta_x \in \mathsf{hom}(\mathcal{M}([0, 1]^k))$ corresponding to all points $x \in [0, 1]^k$ as per Proposition 7.7. In other words, for any $x \in [0, 1]^k$ we must prove

$$s \circ (r_1 - \eta_x(\mathsf{f}_1), \ldots, r_m - \eta_x(\mathsf{f}_m)) = s \circ (r_1 - \mathsf{f}_1(x), \ldots, r_m - \mathsf{f}_m(x)) < 0. \quad (8.4)$$

To this purpose, we make the following

Claim: For any $x \in [0, 1]^k$ there is a $(k + 1)$-element set $H_x \subseteq \mathsf{hom}(\mathcal{M}([0, 1]^k)$ such that the m-tuple $(\eta_x(\mathsf{f}_1), \ldots, \eta_x(\mathsf{f}_m)) \subseteq [0, 1]^m$ given by restricting η_x to $(\mathsf{f}_1, \ldots, \mathsf{f}_m)$ is a convex combination (with coefficients only depending on x) of the restrictions to $(\mathsf{f}_1, \ldots, \mathsf{f}_m)$ of the homomorphisms in H_x.

As a matter of fact, let $T = T_x = \mathsf{conv}(u_0, \ldots, u_k)$ be a k-simplex of ∇ such that $x \in T$. In barycentric coordinates, x can be expressed as

$$x = \sum_{i=0}^{k} \lambda_i u_i \quad \text{for some } \lambda_i \geq 0 \text{ with } \sum_{i=0}^{k} \lambda_i = 1. \quad (8.5)$$

Since for each $j \in \{1, \ldots, m\}$ the function f_j is linear over T, by Proposition 7.7 (7.7) the value $\eta_x(\mathsf{f}_j) = \mathsf{f}_j(x)$ is given by the convex combination of the values $\eta_{u_0}(\mathsf{f}_j) = \mathsf{f}_j(u_0), \ldots, \eta_{u_k}(\mathsf{f}_j) = \mathsf{f}_j(u_k)$ with the same coefficients $\lambda_0, \ldots, \lambda_k$. In symbols,

$$\eta_x(\mathsf{f}_j) = \mathsf{f}_j(x) = \sum_{i=0}^{k} \lambda_i \eta_{u_i}(\mathsf{f}_j) = \sum_{i=0}^{k} \lambda_i \mathsf{f}_j(u_i). \quad (8.6)$$

Our claim is thus settled.

Since the vector $s \in \mathbb{R}^m$ forms an obtuse angle with each vector $d_1, \ldots, d_w$ in (8.3), s also forms an obtuse angle with any convex combination d in $\mathbb{R}^m$ of the vectors d_t, $t = 1, \ldots, w$. In particular, s forms an obtuse angle with the vector $d^* \in \mathbb{R}^m$ given by

$$\begin{aligned} d^* &= \lambda_0(r_1 - \eta_{u_0}(\mathsf{f}_1), \ldots, r_m - \eta_{u_0}(\mathsf{f}_m)) + \cdots + \lambda_k(r_1 - \eta_{u_k}(\mathsf{f}_1), \ldots, r_m - \eta_{u_k}(\mathsf{f}_m)) \\ &= \lambda_0(r_1 - \mathsf{f}_1(u_0), \ldots, r_m - \mathsf{f}_m(u_0)) + \cdots + \lambda_k(r_1 - \mathsf{f}_1(u_k), \ldots, r_m - \mathsf{f}_m(u_k)) \\ &= \text{the } (r_1, \ldots, r_m)\text{-translate of the vector } \left(-\sum_{i=0}^{k} \lambda_i \mathsf{f}_1(u_i), \ldots, -\sum_{i=0}^{k} \lambda_i \mathsf{f}_m(u_i)\right) \\ &= \left(r_1 - \sum_{i=0}^{k} \lambda_i \mathsf{f}_1(u_i), \ldots, r_m - \sum_{i=0}^{k} \lambda_i \mathsf{f}_m(u_i)\right) \\ &= (r_1 - \eta_x(\mathsf{f}_1), \ldots, r_m - \eta_x(\mathsf{f}_m)), \text{ by (110).} \end{aligned}$$

We have just proved that condition (8.4) holds for x and its associated homomorphism

η_x. Since x is an arbitrary point of $[0, 1]^k$, the proof is complete.

□

Conclusion of the proof of Theorem 8.2. The PSAT$_\infty$ problem of checking inconsistency over the infinite set of possible worlds

$$\mathsf{hom}\,\mathcal{M}([0,1]^k) = \{\eta_x \in \mathsf{hom}\,\mathcal{M}([0,1]^k) \mid x \in [0,1]^k\},$$

has been reduced by Lemma 8.9 to checking inconsistency only over the finite set $H = \{\eta_{x_1}, \ldots, \eta_{x_w}\}$. This is the set of homomorphisms corresponding via Proposition 7.7 to the vertices $x_1, \ldots, x_w$ of a triangulation ∇ of $[0, 1]^k$ that linearizes all McNaughton functions coded by the formulas $\psi_1, \ldots, \psi_m$ in the book. The existence and computability of such triangulation ∇ on input $\psi_1, \ldots, \psi_m$ follows from the computability of the linear pieces of all functions $\mathfrak{f}_1, \ldots, \mathfrak{f}_m$ coded by these formulas (Lemma 8.7), together with Corollary 8.6, resting on the constructive character of the proof of Theorem 8.5.

Thus to decide our instance of PSAT$_\infty$ we may restrict to the homomorphisms $\eta_{x_t} \colon \mathcal{M}([0,1]^k) \to [0,1]$, for x_t ranging over the *finite* set $\{x_1, \ldots, x_w\}$ of vertices of ∇. Let the $w \times m$ *rational* matrix M be defined by

$$M_{tj} = r_j - \eta_{x_t}(\mathfrak{f}_j) = r_j - \mathfrak{f}_j(x_t), \quad t = 1, \ldots, w; \quad j = 1, \ldots, m. \tag{8.7}$$

The set of pairs $\{(\mathfrak{f}_1, r_1), \ldots, (\mathfrak{f}_m, r_m)\}$ is *inconsistent* in the sense of Definition 8.1 if and only if some (column) vector $s \in \mathbb{R}^m$ provides a solution to the following system of w strict linear inequalities with rational coefficients:

$$Ms \quad = \underbrace{(z_1, \quad z_2, \quad \ldots \quad, z_w)}_{\text{vector whose coordinates are all } <0}. \tag{8.8}$$

This problem can be decided (in exponential time) by a Turing machine implementing, e.g., Fourier-Motzkin elimination. For a faster decision procedure, Linear programming decides the solvability of the same system in polytime. Inconsistency amounts to showing that a solution s of (8.8) exists in $\mathbb{R}^m$, whence it exists in $\mathbb{Q}^m$. By direct inspection of (8.7) and the Farkas(-Stiemke-Gordan) Lemma 2.6, consistency is equivalent to the existence of a convex combination ς of the homomorphisms $\eta_{x_1}, \ldots, \eta_{x_w}$ such that

$$((\mathfrak{f}_1, \varsigma(\mathfrak{f}_1)), \ldots, ((\mathfrak{f}_m, \varsigma(\mathfrak{f}_m))) = \{(\mathfrak{f}_1, r_1), \ldots, (\mathfrak{f}_m, r_m)\} = \beta.$$

Evidently, ς is a state of $\mathcal{M}([0,1]^k)$ providing the desired extension of β. □

The final part of the proof also yields the following result:

Corollary 8.10 *PSAT$_\infty$ is equivalent to the following problem: Can the set $\beta = \{(\mathfrak{f}_1, r_1), \ldots, (\mathfrak{f}_m, r_m)\}$ be extended to a convex combination of homomorphisms of $\mathcal{M}([0, 1]^k)$ into $[0, 1] \cap \mathbb{Q}$? By Theorem 8.2, also this problem is decidable.* □

Historical and bibliographical notes. McNaughton's 1951 representation theorem for free MV-algebras goes back to [11], seven years before the introduction of MV-algebras. See TheBook, §9.1 for a constructive proof.

For the syntax and semantics of Łukasiewicz logic see TheBook, Sect. 4. Here one can find a proof of the co-NP-completeness of the Ł$_\infty$-tautology problem.

A standard reference for Turing machines, decidability, and related topics is [9]. Also see [3] and [12].

For proofs of both directions of the "Fundamental Theorem 8.5 on Polyhedra" see, e.g., [5, Theorem 1.2] and [2, Theorem 5-10]. For details on the refinement of a polyhedral complex into a triangulation without adding no new vertices see [10, Lemma 1.4 p.11] or [16, 2.9].

Readers interested in implementations of the main algorithms in this chapter may refer to [20, Notes, pp. 47–49, and references therein]. In particular, for complexity estimates of Fourier-Motzkin elimination see [8]. For systems of inequalities in the theory of linear and integer programming see [17, 7-2-7.6] and [15].

A different proof of the decidability of the consistency problem for the free MV-algebra $\mathcal{M}([0, 1]^k)$ was first given in [13] using the Tarski-Seidenberg decision method for elementary algebra, [18, 19]. Later, Bova and Flaminio in [1] gave a sharper estimate. The estimate $w \leq k + 1$ in [13, Corollary 5.4(iv)] does not hold in general. This has no effect on the proof of the decidability result in that paper.

For computational applications see [4, 6, 7, 14].

References

1. Bova, S., & Flaminio, T. (2010). The coherence of Łukasiewicz assessments is NP-complete. *International Journal of Approximate Reasoning, 51*, 294–304.
2. Brønsted, A. (1983). An introduction to convex polytopes, Graduate Texts in Mathematics (Vol. 90). Springer.
3. Davis, M. D., Sigal, R., & Weyuker, E. J. (1994). Computability, complexity, and languages: Fundamentals of theoretical computer science (2nd ed.). Academic Press.
4. De Bona, G., & Finger, M. (2015). Measuring inconsistency in probabilistic logic: Rationality postulates and Dutch book interpretation. *Artificial Intelligence, 227*, 140–164.
5. Ewald, G. (1996). *Combinatorial convexity and algebraic geometry, Graduate Texts in Mathematics* (Vol. 168). New York: Springer.
6. Finger, M., & Preto, S. (2020). Probably partially true: Satisfiability for Łukasiewicz infinitely-valued probabilistic logic and related topics. *Journal of Automated Reasoning, 64*, 1269–1286.
7. Flaminio, T., & Ugolini, S. (2024). Encoding de Finetti's coherence within Łukasiewicz logic and MV-algebras. Annals of Pure and Applied Logic, 175, Article 103337.
8. Jing, R. J., Moreno-Maza, M., Talaashrafi, D. (2020). Complexity estimates for Fourier-Motzkin elimination. In F. Boulier, M. England, T. M. Sadykov, & E. V. Vorozhtsov (Eds.), Computer algebra in scientific computing (Vol. 12291). CASC 2020. Lecture Notes in Computer Science. Springer.

9. Hermes, H. (2012). Enumerability, decidability, computability. An introduction to the theory of recursive functions (2nd ed., Vol. 127). Grundlehren der mathematischen Wissenschaften. Springer.
10. Hudson, J. F. P. (1969). *Piecewise linear topology*. W.A: Benjamin Inc.
11. McNaughton, R. (1951). A theorem about infinite-valued sentential logic. *Journal of Symbolic Logic, 16*, 1–13.
12. Monk, J. D. (1976). Mathematical logic, Graduate Texts in Mathematics (Vol. 37). Springer.
13. Mundici, D. (2006). Bookmaking over infinite-valued events. *International Journal of Approximate Reasoning, 43*, 223–240.
14. Preto, S., & Finger, M. (2023). Proving properties of binary classification neural networks via Łukasiewicz logic. *Logic Journal of the IGPL, 31*(5), 805–821.
15. Roos, C., Terlaky, T., & Vial, J.-P. (2006). Interior point methods for linear optimization (2nd ed.). Springer.
16. Rourke, C. P., & Sanderson, B. J. (1972). *Introduction to piecewise-linear topology*. Springer.
17. Schrijver, A. (1986). *Theory of linear and integer programming*. Wiley.
18. Seidenberg, A. (1954). A new decision method for elementary algebra. *Annals of Mathematics, 60*, 365–374.
19. Tarski, A. (1951). *A decision method for elementary algebra and geometry*. University of California Press.
20. Ziegler, G. M. (1995). Lectures on polytopes, Graduate Text in Mathematics, (Vol. 152). Springer.

Chapter 9
Deciding the Conditional PSAT_∞ Problem

Conditionals have a pervasive role in probability theory. This chapter extends the PSAT_∞ problem to situations where events are subject to a finite set Θ of conditions. Their effect is to restrict the set of "possible worlds" (= homomorphisms into [0, 1] of the ambient MV-algebra of events) to the set of possible worlds where all conditions in Θ hold true. Each condition in Θ is definable by a formula in Łukasiewicz logic. Extending the results of the foregoing chapter, we will prove the decidability of the resulting "conditional" PSAT_∞ problem.

9.1 Conditioning [0, 1]-Valued Continuous Events

Let $X_1, \dots, X_k$ be the first k variables in the syntax of Łukasiewicz logic. Any two variables X_i, X_j, $(i \neq j)$ code events which are as unrelated as possible. Correspondingly, (as shown in Chapter 1, §1.4), the set of "possible worlds" W assigning a truth-value to the events coded by the variables $X_1, \dots, X_k$ is as large as possible. Formally, W is identified with the set of all valuations of k-variable formulas, as defined in 9.3 below.

Providing additional information about the mutual relations between the variables amounts to restricting the set W, in analogy with the example in Chapter 1, §1.4. To fix ideas, let us assume that the variables are subject to the conditions coded by the following formulas $\theta_1, \theta_2, \theta_3$:

(θ_1): $\neg X_1 \oplus \neg X_2$, stating that X_1 implies $\neg X_2$;
(θ_2): $\neg X_2 \oplus X_3$, stating that X_2 implies X_3;
(θ_3): $X_1 \oplus X_4 \oplus X_5$, stating that their disjunction is true in Łukasiewicz logic.

The combined effect of the conditions coded by the set $\Theta = \{\theta_1, \theta_2, \theta_3\}$ is that the set of possible worlds for $X_1, \dots, X_5$ is restricted to those valuations assigning the truth-value 1 to each formula in Θ. For instance, condition θ_2 restricts to those

D. Mundici, *Algebraic Probabilistic Consistency*, Trends in Logic 69,
https://doi.org/10.1007/978-3-031-98337-5_9

valuations U with $U(X_2) \leq U(X_3)$; condition θ_3 restricts to those valuations V such that $V(X_1) + V(X_4) + V(X_5) \geq 1$. Each condition in Θ must be considered 100% true. Since the set Θ is finite, the conjunction

$$(\neg X_1 \oplus \neg X_2) \odot (\neg X_2 \oplus X_3) \odot (X_1 \oplus X_4 \oplus X_5)$$

is a single formula θ having the same conditioning effect as the set Θ of the three formulas in Θ. Accordingly, we will always reformulate any finite set Θ of conditions in Łukasiewicz logic as a singleton $\{\theta\} = \{\theta_\Theta\}$.

Generalizing the above example, the variables X_i may be replaced by k-variable Ł$_\infty$-formulas $\psi_1, \ldots, \psi_m$ coding more complex events, as in the previous chapter. The finite set Θ of conditions may be thought of as consisting of two parts: a first part of formulas of the form $\theta'_1(\psi_1, \ldots, \psi_m), \ldots, \theta'_{k'}(\psi_1, \ldots, \psi_m)$ relating the events described by $\psi_1, \ldots, \psi_m$; and a second part consisting of formulas $\theta''_1(\psi_1, \ldots, \psi_m), \ldots, \theta''_{k''}(\psi_1, \ldots, \psi_m)$ imposing further conditions on $\psi_1, \ldots, \psi_m$. At the end of the day, the conditioning set $\Theta' \cup \Theta''$ for $\psi_1, \ldots, \psi_m$ again boils down to a singleton formula $\{\theta\}$.

Summing up, the input of our conditional decision problem is given by a finite set of events coded by k-variable formulas $\psi_1, \ldots, \psi_m$ subject to a single condition, also coded by a k-variable formula θ. In other words, the input of this problem is a *string of symbols*

$$\{\psi_1, \ldots, \psi_m;\ \theta\}, \text{ with } r_j \in \mathbb{Q} \cap [0, 1] \text{ for each Ł}_\infty\text{-formula } \psi_j. \tag{9.1}$$

To attract bettors, Bookie adds informal descriptions of the intended meaning of all these formulas. But all computations concerning the consistency of this conditioned book only depend on the input list of symbols in (9.1), and so do all financial transactions between the ideal bookmaker Bookie and all bettors willing to buy/sell bets with his betting odds r_j.

9.2 The Conditional PSAT$_\infty$ Problem

Definition 9.1 The *conditional PSAT$_\infty$ problem* is defined as follows:
INPUT: A list $\{(\psi_1, r_1), \ldots, (\psi_m, r_m);\ \theta\}$ of k-variable Ł$_\infty$-formulas, together with rational numbers $r_1, \ldots, r_m \in [0, 1]$ written in binary notation.
QUESTION: Letting $\beta = \{(\mathsf{f}_{\psi_1} \restriction \mathsf{Mod}\,\theta,\ r_1), \ldots, (\mathsf{f}_{\psi_m} \restriction \mathsf{Mod}\,\theta,\ r_m)\}$ and $\mathsf{Mod}\,\theta = \{x \in [0, 1]^k \mid \mathsf{f}_\theta(x) = 1\}$, is β *consistent* in the MV-algebra $\mathcal{M}([0, 1]^k) \restriction \mathsf{Mod}\,\theta$?

Corollary 9.13 will provide various equivalent reformulations of the conditional PSAT$_\infty$ problem.

Theorem 9.2 *The conditional PSAT$_\infty$ problem is decidable.*

The remainder of this chapter is devoted to proving this result.

Outline of the proof.

The semantic and syntactic properties of "consequence" in Łukasiewicz logic will provide the basic tools to prove the Turing decidability of the conditional PSAT$_\infty$ problem. The set of "possible worlds" will be understood as the set of valuations satisfying the conditioning formula θ. For our problem to make sense we must preliminarily check that θ is satisfiable, (i.e., logically consistent) i.e., $V(\theta) = 1$ for at least one valuation V.

The logical equivalence of two formulas ϕ and ψ in Łukasiewicz logic is generalized to logical equivalence $\equiv_\theta$ "subject to" θ by the following stipulation: $\psi \equiv_\theta \phi$ if $V(\psi) = V(\phi)$ for every valuation V satisfying θ. MV-algebras consisting of $\equiv_\theta$ classes of formulas constitute the most general *finitely presented* MV-algebra, i.e., a quotient of a finitely generated free MV-algebra by a *principal* ideal, i.e., an ideal generated by an element of the algebra. See TheBook p. 13 for details.

Proving the conditional consistency problem then amounts to giving a constructive refinement of the proof of Theorem 8.2, uniformly over any finite presentation $\{(\psi_1, r_1), \ldots, (\psi_m, r_m);\ \theta\}$. As a first step, the MV-algebra A is identified with the MV-algebra of restrictions of the functions of $\mathcal{M}([0,1]^k)$ to $\mathsf{Mod}\,\theta = \bigcup \nabla = |\nabla|$, for ∇ a computable rational triangulation over whose simplexes each function f_{ψ_j} is linear.

The equivalence between quotients of free MV-algebras $\mathcal{M}([0,1]^k)$ by principal ideals and restrictions of the functions in $\mathcal{M}([0,1]^k)$ to supports of rational triangulations of $[0,1]^k$ will be proved in Proposition 9.9(ii).

At the end of the day, checking the inconsistency of β will be reduced to checking it over the finite set of vertices of a rational triangulation ∇ of $\mathsf{Mod}\,\theta$ such that every function $\mathsf{f}_{\psi_1}, \ldots, \mathsf{f}_{\psi_m}$ is linear over each simplex of ∇. This requires a finite number of Turing steps.

9.3 Valuations, Formulas, Logical Equivalence, Consequence

The prefix "Ł$_\infty$-" for formulas, tautologies, valuations, syntactic and semantic consequence will be omitted whenever there is no danger of confusion.

For $k = 1, 2, \ldots$ we will adopt the notation

$$\mathsf{Form}_k = \{\phi \mid \phi \text{ is a formula in the variables } X_1, \ldots, X_k\}.$$

Following standard usage we allow the possibility that not all variables in the set $\{X_1, \ldots, X_k\}$ occur in ϕ. Anyway, from the input formulas $\psi_1, \ldots, \psi_m, \theta$ of our conditional consistency problem, the parameter k will be unambiguously computed from the set of all variables occurring in the input formulas.

Form_k is also known as the *absolutely free algebra* of all k-variable formulas. In TheBook, p. 52 one finds the notation $\mathsf{Term}_k^{[0,1]}$ for our present Form_k.

Definition 9.3 A *valuation* is a function $V : \mathsf{Form}_k \to [0, 1]$ such that

$$V(\neg\phi) = 1 - V(\phi) \quad \text{and} \quad V(\phi \oplus \psi) = \min(1, V(\phi) + V(\psi)).$$

Our valuations are said to be "[0, 1]-valuations" in TheBook, 4.1.2. Any valuation $V : \mathsf{Form}_k \to [0, 1]$, is *uniquely* determined by the values it assigns to the variables. This follows from the unique readability of formulas. (See TheBook, pp. 21–22 for details.)

A formula θ is *satisfiable* (= *logically consistent*) if $V(\theta) = 1$ for some valuation V. We also say that θ is a *semantic tautology* if $V(\theta) = 1$ for every valuation V. By a *syntactic tautology* (called a "provable formula" in TheBook.4.3.3.) we mean a formula which is obtainable from the MV-algebraic axioms via finitely many applications of substitution and Modus Ponens. See TheBook §4.3 for several detailed examples of derivations of syntactic tautologies in Łukasiewicz logic.

Two formulas $\phi, \psi \in \mathsf{Form}_k$ are said to be *semantically equivalent* if $V(\phi) = V(\psi)$ for any valuation V. On the other hand, ϕ and ψ are *syntactically equivalent* if $\phi \leftrightarrow \psi$ is a tautology.

Proposition 9.4 *Two formulas are semantically equivalent if and only if they are syntactically equivalent. Thus in particular, syntactic and semantic tautologies coincide.*

Proof See TheBook, 4.5.2. □

We write $\phi \equiv \psi$ if ψ and ϕ are (syntactically=semantically) equivalent. The $\equiv$-equivalence class of ϕ is denoted $\phi/\equiv$. In TheBook, p. 93 one finds the notation $|\phi|$ for $\phi/\equiv$.

With the MV-algebraic operations naturally induced by the connectives, the set $\mathsf{Form}_k/\equiv$ of equivalence classes of formulas becomes an MV-algebra.

Consequence in Łukasiewicz logic. For any set Θ of formulas and formula ϕ we say that ϕ is a *syntactic consequence of* Θ, and write

$$\Theta \vdash \phi,$$

if ϕ is obtainable from some (necessarily finite) subset of Θ and the tautologies by finitely many applications of Modus Ponens. Of course, ϕ is a tautology iff $\emptyset \vdash \phi$ iff $\{X_1 \leftrightarrow X_1\} \vdash \phi$. See TheBook §4.3 for detailed examples of derivations of ϕ from Θ.

We say that ϕ is a *(tarskian) semantic consequence* of Θ in Łukasiewicz logic,

$$\Theta \models \phi,$$

if every valuation satisfying all formulas in Θ also satisfies ϕ.

Theorem 9.5 *(i) When Θ is finite, $\Theta \vdash \phi$ if and only if $\Theta \models \phi$.*
(ii) This no longer holds in general when Θ is infinite.

Proof (i) The equivalence of $\vdash$ and $\models$ for any finite set Θ of premises is proved in TheBook, 4.6.7.

(ii) A counterexample of this equivalence for infinite Θ exists already for $\Theta \subseteq \mathsf{Form}_1$ as follows: For $n = 1, 2, \ldots$ let $f_n \colon [0, 1] \to [0, 1]$ be defined by $f_n(x) = \max(0, \min(1, 2 - nx))$. Then $f_n(x) = 1$ iff $x \in [0, 1/n]$. The graph of the continuous function f_n has three consecutive linear pieces, respectively represented by the linear polynomials $1,\ 2 - nx,\ 0$. By direct inspection, f_n is a McNaughton function, $f_n \in \mathcal{M}([0, 1])$. Let θ_n be a one-variable formula such that $\mathsf{f}_{\theta_n} = f_n$. By definition of semantic consequence, the set $\Theta = \{\theta_1, \theta_2, \ldots\}$ satisfies the condition $\Theta \models \neg X$. On the other hand, $\Theta \not\vdash \neg X$. To see this, arguing by way of contradiction, let us suppose $\Theta \vdash \neg X$. By definition of syntactic consequence, for some finite subset Θ' of Θ we have $\Theta' \vdash \neg X$. Then, by the first part of the proof, $\Theta' \models \neg X$, i.e., all valuations satisfying Θ' also satisfy $\neg X$. This contradicts the fact that $\mathsf{f}_{\neg X}$ takes value 1 only at $x = 0$, while there is $0 < \epsilon \in \mathbb{Q}$ such that for each $\psi \in \Theta'$ the McNaughton function f_ψ takes value 1 over the larger set $[0, \epsilon]$. □

Remark 9.6 In the final section on historical and bibliographical notes we will briefly describe a non-tarskian, differential semantic notion $\models_\partial$ of consequence in Łukasiewicz logic that takes into account the differential structure of McNaughton functions and avoids the phenomenon described in the second part of the above proof, resulting in the discrepancy between $\models$ and $\vdash$.

Throughout the remainder of this chapter, Θ will stand for a fixed but otherwise arbitrary finite set of premises, coded by the single formula θ given by the conjunction of the formulas in Θ. While we will mostly write $\theta \vdash \phi$, by Theorem 9.5(i) it is understood that this is equivalent to $\theta \models \phi$. We will freely say "ϕ is a consequence of θ" without specifying whether we mean syntactic or semantic consequence.

Unless we specify otherwise, θ will henceforth denote a satisfiable formula.

Proposition 9.7 *$\theta \vdash \phi$ if and only if for some $n = 1, 2, \ldots$ the formula $\underbrace{\theta \odot \cdots \odot \theta}_{n \text{ times}} \to \phi$ is a tautology.*

Proof TheBook, 4.6.5. □

Proposition 9.8 *For any $k = 1, 2, \ldots$ we have:*

(i) $\mathsf{Form}_k/\equiv$ is (an isomorphic copy of) the free MV-algebra on the free generating set $\{X_1/\equiv, \ldots, X_k/\equiv\}$.

(ii) Upon writing $\upsilon(X_i/\equiv) = V(X_i)$, for each $i = 1, \ldots, k$, every valuation V determines a unique homomorphism $\upsilon_V \in \mathsf{hom}(\mathsf{Form}_k/\equiv)$ of the MV-algebra $\mathsf{Form}_k/\equiv$ into the standard MV-algebra $[0, 1]$.

(iii) For every $\upsilon \in \mathsf{hom}(\mathsf{Form}_k/\equiv)$ *let* $V = V_\upsilon$ *be the unique valuation of* Form_k *with* $V(X_i) = \upsilon(X_i/\equiv)$ *for each* $i = 1, \ldots, k$. *Then* $\upsilon = V_\upsilon$.

(iv) The map $X_i/\!\models\, \in \mathsf{Form}_k/\!\models\, \mapsto \pi_i \in \mathcal{M}([0,1]^k)$ *uniquely extends to an isomorphism*

$$\omega\colon \phi/\equiv\ \in \mathsf{Form}_k/\equiv\ \cong\ \mathsf{f}_\phi \in \mathcal{M}([0,1]^k).$$

The inverse of ω *is the unique homomorphism of* $\mathcal{M}([0,1]^k)$ *into* $\mathsf{Form_k}/\equiv$ *which extends the map* $\pi_i \mapsto X_i/\equiv$.

(v) The map $\mathfrak{J} \mapsto \omega(\mathfrak{J})$ *is a lattice-isomorphism of the set of ideals of* $\mathsf{Form}_k/\equiv$ *onto the set of ideals of* $\mathcal{M}([0,1]^k)$, *both sets being ordered by inclusion. We have the isomorphism*

$$(\mathsf{Form_k}/\equiv)/\mathfrak{J}\ \cong\ \mathcal{M}([0,1]^k)/\omega(\mathfrak{J}). \tag{9.2}$$

(vi) The map $x \mapsto \upsilon_x$, *where* $\upsilon_x(X_i/\equiv) = x_i$, *yields a bijection of* $[0,1]^k$ *onto* $\mathsf{hom}(\mathsf{Form}_k/\equiv)$.

Proof (i) See TheBook, 3.1.4.

(ii) Every valuation V gives the same value to any two $\equiv$-equivalent formulas. Now use the fact that $\{X_1/\equiv, \ldots, X_k/\equiv\}$ freely generates $\mathsf{Form_k}/\equiv$. Compare with TheBook, 4.5.6.

(iii) By direct inspection.

(iv) This follows from (i) and Theorem 7.8, because $\{X_1/\equiv, \ldots, X_k/\equiv\}$ and $\{\pi_1, \ldots, \pi_k\}$ are free generating sets in their respective free MV-algebras $\mathsf{Form_k}/\equiv$ and $\mathcal{M}([0,1]^k)$.

(v) Immediate from (iv).

(vi) Combine (iv) with Proposition 7.7. Observe that, by (i), the homomorphism υ_x is uniquely determined by its restriction to $\{X_1/\equiv, \ldots, X_k/\equiv\}$. □

9.4 Preparing Equivalent Versions of the Conditional PSAT$_\infty$ Problem

Lindenbaum algebra, θ-equivalence, ideals and consequence. Fix $k = 1, 2, \ldots$. For $\phi, \psi, \theta \in \mathsf{Form}_k$ we write

$$\phi \equiv_\theta \psi \quad \text{iff} \quad \theta \vdash \phi \leftrightarrow \psi \quad \text{iff} \quad \theta \models \phi \leftrightarrow \psi, \tag{9.3}$$

and say that ϕ and ψ are *θ-equivalent.* Likewise, two equivalence classes $\phi/\equiv$ and $\psi/\equiv$ are said to be $\equiv_\theta$-equivalent if $\phi \equiv_\theta \psi$. For each $\phi \in \mathsf{Form}_k$ we let $\phi/\equiv_\theta$ denote the $\equiv_\theta$-equivalence class of ϕ. When the context is clear, $\phi/\equiv_\theta$ may be naturally identified with the equivalence class $(\phi/\equiv)/\equiv_\theta$. The set

$$\mathsf{Form}_k/\equiv_\theta\ =\ (\mathsf{Form}_k/\equiv)/\equiv_\theta$$

of $\equiv_\theta$-equivalence classes of formulas, equipped with the MV-algebraic operations naturally induced by the connectives, is called the *Lindenbaum algebra* of θ. Compare with TheBook, §§4.4-4.6 where a different notation is used.

Recall that all ideals in this paper are assumed to be *proper*. Hence, for every principal ideal $\mathfrak{i}$ of $\mathcal{M}([0,1]^k)$ there is a satisfiable formula $\theta \in \mathsf{Form}_k$ such that $\mathfrak{i}$ is the ideal $\mathfrak{i}_{\neg\theta}$ generated by the McNaughton function $\mathsf{f}_{\neg\theta}$,

$$\mathfrak{i}_{\neg\theta} = \text{ ideal generated by } \mathsf{f}_{\neg\theta}, \quad (Z\mathsf{f}_{\neg\theta} \text{ being nonempty}). \tag{9.4}$$

In TheBook, $\mathfrak{i}_{\neg\theta}$ is denoted $\langle\neg\theta\rangle$. Let the set $\mathsf{Mod}\,\theta \subseteq [0,1]^k$ be defined as follows:

$$\mathsf{Mod}\,\theta = Z\mathsf{f}_{\neg\theta} = \{x \in [0,1]^k \mid \mathsf{f}_\theta(x) = 1\} = \mathsf{f}_\theta^{-1}(1).$$

As the zeroset of a McNaughton function, $\mathsf{Mod}\,\theta$ is a finite union U of rational simplexes in $[0,1]^k$. Furthermore, an inessential modification of the Turing machine $\mathcal{T}_{\text{linearizer}}$ of Lemma 8.7 explicitly computes a rational triangulation of U. Note that $\mathsf{Mod}\,\theta$ need not be a union of k-dimensional simplexes, as it might contain isolated points, tails consisting of rational segments, and the like.

Proposition 9.9 *With the principal ideal* $\mathfrak{i}_{\neg\theta}$ *defined in* (9.4) *above, we have:*

(i) A McNaughton function $g \in \mathcal{M}([0,1]^k)$ *belongs to* $\mathfrak{i}_{\neg\theta}$ *if and only if its zeroset* Zg *contains the zeroset* $Z\mathsf{f}_{\neg\theta} = \mathsf{Mod}\,\theta$.

(ii) The map $\rho_\theta\colon g/\mathfrak{i}_{\neg\theta} \mapsto g \restriction \mathsf{Mod}\,\theta$ *is an isomorphism of* $\mathcal{M}([0,1]^k)/\mathfrak{i}_{\neg\theta}$ *onto* $\mathcal{M}([0,1]^k) \restriction \mathsf{Mod}\,\theta = \{g \restriction \mathsf{Mod}\,\theta \mid g \in \mathcal{M}([0,1]^k)\}$,

$$\rho_\theta\colon \mathcal{M}([0,1]^k)/\mathfrak{i}_{\neg\theta} \cong \mathcal{M}([0,1]^k) \restriction \mathsf{Mod}\,\theta. \tag{9.5}$$

Proof (i) This is proved in TheBook 3.4.8.

(ii) Following TheBook 1.2.4 (where the notation "d" is used for Chang's distance function), for any MV-algebra A and elements $a, b \in A$ let

$$\mathsf{distance}(a,b) = (a \ominus b) \oplus (b \ominus a).$$

In the standard MV-algebra $[0,1]$, $\mathsf{distance}(x,y) = |x-y|$. Thus for any $\phi, \psi \in \mathsf{Form}_k$, $(\mathsf{distance}(\mathsf{f}_\phi, \mathsf{f}_\psi))(x) = |\mathsf{f}_\phi(x) - \mathsf{f}_\psi(x)|$ for all $x \in [0,1]^k$. We then have:

$$\begin{aligned}
\mathsf{f}_\phi/\mathfrak{i}_{\neg\theta} = \mathsf{f}_\psi/\mathfrak{i}_{\neg\theta} &\Leftrightarrow \mathsf{distance}(\mathsf{f}_\phi, \mathsf{f}_\psi) \in \mathfrak{i}_{\neg\theta} \\
&\Leftrightarrow Z\,\mathsf{distance}(\mathsf{f}_\phi, \mathsf{f}_\psi) \supseteq Z\mathsf{f}_{\neg\theta}, \quad \text{by (i)} \\
&\Leftrightarrow \text{for all } x \in [0,1]^k, \quad (\mathsf{f}_{\neg\theta}(x) = 0 \Rightarrow \mathsf{distance}(\mathsf{f}_\phi(x), \mathsf{f}_\psi(x)) = 0) \\
&\Leftrightarrow Z\mathsf{f}_{\neg\theta} \subseteq \{x \in [0,1]^k \mid \mathsf{f}_\psi(x) = \mathsf{f}_\phi(x)\} \\
&\Leftrightarrow \mathsf{f}_\psi = \mathsf{f}_\phi \text{ over } \mathsf{Mod}\,\theta \\
&\Leftrightarrow \mathsf{f}_\psi \restriction \mathsf{Mod}\,\theta = \mathsf{f}_\phi \restriction \mathsf{Mod}\,\theta.
\end{aligned}$$

Therefore, ρ_θ is a bijection of $\mathcal{M}([0,1]^k)/\mathfrak{i}_\theta$ onto $\mathcal{M}([0,1]^k) \restriction \mathsf{Mod}\,\theta$. The rest easily follows by direct inspection. □

We next consider the ideal $\mathfrak{J}_{\neg\theta}$ of $\mathsf{Form}_k/\equiv$ corresponding to the ideal $\mathfrak{i}_{\neg\theta}$ of $\mathcal{M}([0,1]^k)$ via Proposition 9.8(v). By (9.2),

$$(\mathsf{Form}_k/\equiv)/\mathfrak{J}_{\neg\theta} \cong \mathcal{M}([0,1]^k)/\mathfrak{i}_{\neg\theta}. \tag{9.6}$$

Since $\mathfrak{i}_{\neg\theta}$ is the principal ideal of $\mathcal{M}([0,1]^k)$ generated by $\mathsf{f}_{\neg\theta}$ then, by Proposition 9.8(iv), $\mathfrak{J}_{\neg\theta}$ is the principal ideal of $\mathsf{Form}_k/\equiv$ generated by $\neg\theta/\equiv$.

A useful characterization of $\mathfrak{J}_{\neg\theta}$ in terms of syntactic consequence is given by the following result:

Proposition 9.10 *With ω as in Proposition 9.8(iv)–(v), let $\mathfrak{J}_{\neg\theta}$ be the ideal of $\mathsf{Form}_k/\equiv$ generated by $\neg\theta/\equiv$. Then*

$$\omega^{-1}(\mathfrak{i}_{\neg\theta}) = \mathfrak{J}_{\neg\theta} = \{\phi/\equiv \mid \theta \vdash \neg\phi\}.$$

Proof *First part: For any $\phi \in \mathsf{Form}_k$, $\phi/\equiv\ \in \mathfrak{J}_{\neg\theta} \ \Leftrightarrow\ \theta \vdash \neg\phi$.*

$$\begin{aligned}
\phi/\equiv\ \in \mathfrak{J}_{\neg\theta} &\Leftrightarrow \text{for some } n = 1,2,\ldots,\ \ \phi/\equiv\ \le\ n.\neg\theta/\equiv \\
&\Leftrightarrow \text{for some } n,\ \neg\phi/\equiv\ \ge \neg n.\neg\theta/\equiv \\
&\Leftrightarrow \text{for some } n,\ \theta/\equiv\ \odot\cdots\odot\ \theta/\equiv\ (n \text{ times})\ \ \le \neg\phi/\equiv \\
&\Leftrightarrow \text{for some } n,\ ((\theta/\equiv\ \odot\cdots\odot\ \theta/\equiv) \to \neg\phi/\equiv)\ \ \equiv\ (X \oplus \neg X)/\equiv) \\
&\Leftrightarrow \theta \odot \cdots \odot \theta \to \neg\phi \text{ is a tautology} \\
&\Leftrightarrow \theta \vdash \neg\phi, \ \ \text{by Proposition 9.7.}
\end{aligned}$$

Second part: For any $\phi \in \mathsf{Form}_k$, $\theta \vdash \neg\phi \Leftrightarrow \mathsf{f}_\phi \in \mathfrak{i}_{\neg\theta}$.

$$\begin{aligned}
\theta \vdash \neg\phi &\Leftrightarrow \theta \models \neg\phi \text{ by Theorem 9.5(i)} \\
&\Leftrightarrow \text{for all valuations } V, \text{if } \ V(\theta) = 1 \text{ then } V(\phi) = 0 \\
&\Leftrightarrow \text{for all } \upsilon \in \mathsf{hom}(\mathsf{Form}_k/\equiv),\ \text{if } \ \upsilon(\theta/\equiv) = 1 \text{ then } \upsilon(\phi/\equiv) = 0, \\
&\qquad \text{by Proposition 9.8 (ii)} \\
&\Leftrightarrow \text{for all } \eta \in \mathsf{hom}(\mathcal{M}([0,1]^k)),\ \ \text{if } \eta(\mathsf{f}_\theta) = 1 \ \text{then } \eta(\mathsf{f}_\phi) = 0, \\
&\qquad \text{by Proposition 9.8(ii)–(iii)} \\
&\Leftrightarrow \text{for all } x \in [0,1]^k,\ \ \text{if } \mathsf{f}_\theta(x) = 1 \ \text{then } \mathsf{f}_\phi(x) = 0, \text{ by Proposition 7.7} \\
&\Leftrightarrow \mathsf{Zf}_\phi \supseteq \mathsf{Zf}_{\neg\theta} = \mathsf{Mod}\,\theta \\
&\Leftrightarrow \mathsf{f}_\phi \in \mathfrak{i}_{\neg\theta}, \text{ by Proposition 9.9(i).}
\end{aligned}$$

This completes the proof. □

Proposition 9.11 *(i) The map $\nu_\theta : \phi/\equiv_\theta\ \mapsto\ \mathsf{f}_\phi/\mathfrak{i}_{\neg\theta}$ is an isomorphism of the MV-algebra $\mathsf{Form}_k/\equiv_\theta$ onto $\mathcal{M}([0,1]^k) \upharpoonright \mathsf{Mod}\,\theta$,*

$$\nu_\theta : \mathsf{Form}_k/\equiv_\theta\ \cong\ \mathcal{M}([0,1]^k) \upharpoonright \mathsf{Mod}\,\theta.$$

In other words, up to isomorphism, letting θ range over all satisfiable formulas, $\mathsf{Form}_k/{\equiv_\theta}$ *is the most general possible* finitely presented *MV-algebra.*

(ii) In particular, let A be the Lindenbaum algebra of a conjunction Θ of a finite set of formulas. If Θ is satisfiable (in the sense that there is a valuation assigning the value 1 to all conjuncts of Θ) then A is semisimple.

Proof (i) Combine (9.5) and (9.6). Compare with TheBook, 4.5.5, 4.6.9.

(ii) From (i) and Theorem 7.4. □

Remark 9.12 Let $\phi, \theta \in \mathsf{Form}_k$, with θ satisfiable. We have shown: $\theta \vdash \phi$ iff $\theta \models \phi$ iff $\mathsf{f}_\phi(x) = 1$ for all x such that $\mathsf{f}_\theta(x) = 1$ iff $\mathsf{f}_{\neg\phi}$ belongs to the ideal of $\mathcal{M}([0,1]^k)$ generated by $\mathsf{f}_{\neg\theta}$ iff $Z\mathsf{f}_{\neg\psi} \supseteq Z\mathsf{f}_{\neg\theta}$.

When θ is a tautology we have $\mathsf{Form}_k/\equiv_\theta \; = \; \mathsf{Form}_k/\equiv$ = the free n-generated MV-algebra of equivalence classes of k-variable formulas.

With θ satisfiable, Theorem 6.3 combined with the above results concerning Ł$_\infty$, yield the following equivalent reformulations of the conditional PSAT$_\infty$ problem:

Corollary 9.13 *Let $\{(\psi_1, r_1), \ldots, (\psi_m, r_m), \theta\}$ be a set where θ and the ψ_j are k-variable formulas, and $r_j \in [0,1] \cap \mathbb{Q}$ for all $j = 1, \ldots, m$. Then the following conditions are equivalent to giving a positive answer to the conditional PSAT$_\infty$ problem on this instance:*

(i) The set of pairs $\{(\mathsf{f}_{\psi_1} \upharpoonright \mathsf{Mod}\,\theta, r_1), \ldots, (\mathsf{f}_{\psi_m} \upharpoonright \mathsf{Mod}\,\theta, r_m)\}$ can be extended to a state of the MV-algebra $\mathcal{M}([0,1]^k) \upharpoonright \mathsf{Mod}\,\theta$.

(ii) The set $\{(\psi_1/{\equiv_\theta}, r_1) \ldots, (\psi_m/{\equiv_\theta}, r_m)\}$ can be extended to a state of $\mathsf{Form}_k/\equiv_\theta$.

(iii) For any $s = (s_1, \ldots, s_m) \in \mathbb{R}^m$ there is $\eta \in \mathsf{hom}(\mathcal{M}([0,1]^k) \upharpoonright \mathsf{Mod}\,\theta)$ such that $\sum_{j=1}^m s_j(r_j - \eta(\mathsf{f}_{\psi_j} \upharpoonright \mathsf{Mod}\,\theta)) \geq 0$.

(iv) For any $s = (s_1, \ldots, s_m) \in \mathbb{R}^m$ there is $\eta \in \mathsf{hom}(\mathsf{Form}_k/\equiv_\theta)$ such that $\sum_{j=1}^m s_j(r_j - \eta(\psi_j/\equiv_\theta)) \geq 0$.

(v) For any $s = (s_1, \ldots, s_m) \in \mathbb{R}^m$, $\sum_{j=1}^m s_j(r_j - \eta_x(\mathsf{f}_{\psi_j} \upharpoonright \mathsf{Mod}\,\theta)) \geq 0$ for some $x \in \mathsf{Mod}\,\theta$.

(vi) For any $s = (s_1, \ldots, s_m) \in \mathbb{R}^m$, $\sum_{j=1}^m s_j(r_j - \mathsf{f}_{\psi_j}(x)) \geq 0$ for some $x \in$ $\mathsf{Mod}\,\theta$. □

Adopting the equivalent formulation (vi) of the conditional PSAT$_\infty$ we will prove the decidability of the following problem:

INPUT: k-variable formulas $\psi_1, \ldots, \psi_m$ and θ, with rational numbers $r_1, \ldots, r_m \in [0,1]$ written in binary notation, and θ satisfiable.
QUESTION: Does every $s \in \mathbb{R}^m$ yield an $x \in \mathsf{Mod}\,\theta$ with $\sum_{j=1}^m s_j(r_j - \mathsf{f}_{\psi_j}(x)) \geq 0$?

9.5 Proving the Decidability of the Conditional PSAT$_\infty$ Problem

For each $k = 1, 2, \dots$ the following sets are tacitly equipped with a fixed but otherwise arbitrary lexicographic order:

Points in $\mathbb{Q}^k$ (represented by pairs of integers in binary notation);
Rational m-simplexes in $[0, 1]^k$, $m = 0, \dots, k$;
Linear k-variable polynomials with integer coefficients;
Rational hyperplanes in $\mathbb{R}^k$.

We also assume all formulas in Form_k to be lexicographically listed:

$$\phi_{k,1} <_{\text{lex}} \phi_{k,2} <_{\text{lex}} \cdots.$$

Correspondingly, the McNaughton functions in $\mathcal{M}([0, 1]^k)$ can be listed as follows:

$$\mathsf{f}_{k,1}, \mathsf{f}_{k,2}, \dots, \quad \text{with } \mathsf{f}_{k,j} \text{ shorthand for } \mathsf{f}_{\phi_{k,j}}.$$

While in the previous sections the conditioning formula θ is always assumed to satisfy the condition $Z\mathsf{f}_{\neg\theta} \neq \emptyset$, for our computational purposes in this section we must preliminarily check that the input formula θ does satisfy this condition.

The following Turing machine takes care of the problem:

Proposition 9.14 *There is a Turing machine $T_{\mathsf{check\,proper\,ideal}}$ which, on any input formula $\theta \in \mathsf{Form}_k$, decides each one of the following equivalent*[1] *problems:*

(i) Whether the ideal $\mathfrak{i}_{\neg\theta}$ of $\mathcal{M}([0, 1]^k)$ generated by $\mathsf{f}_{\neg\theta}$ is proper.
(ii) Whether $\mathfrak{i}_{\neg\theta} \neq \mathcal{M}([0, 1]^k)$.
(iii) Whether $\mathsf{f}_{\neg\theta}$ has a nonempty zeroset.
(iv) Whether at least one valuation gives 1 to θ (i.e., θ is satisfiable/logically consistent).

Proof We assume the grammatical correctness of θ has already been checked by a parsing algorithm. In view of the results of the previous section, it is sufficient (and necessary) to show the decidability of problem (iii), asking whether $\mathsf{f}_{\neg\theta}$ has an empty zeroset.

Working inductively on the number of connectives in the subformulas of θ, $T_{\mathsf{check\,proper\,ideal}}$ first lists all linear pieces of the McNaughton function $\mathsf{f}_{\neg\theta}$. We now observe that condition (iii) holds if and only if there is a system $\mathcal{L}$ of k linear equations with integer coefficients in the unknowns $x_1, \dots, x_k$ given by the linear pieces of $\mathsf{f}_{\neg\theta}$ such that $\mathsf{f}_{\neg\theta}$ vanishes at the unique solution $v^* \in [0, 1]^k$ of $\mathcal{L}$. Each linear function in $\mathcal{L}$ is a polynomial representing a linear piece of $\mathsf{f}_{\neg\theta}$. Induction on the number of occurrences of connectives in θ shows that the absolute value of each coefficient in $\mathcal{L}$ does not exceed the length $|\theta|$ of θ.

[1] By Corollary 9.13.

Hadamard's theorem on determinants yields a polynomial $p_1 \colon \mathbb{R} \to \mathbb{R}$ such that the denominator d of v^* satisfies the inequality $d \le p_1(|\theta|)$. This polynomial only depends on the parameter k. The fact that its degree increases with k does not impinge on the decidability of our problem. There is a polynomial $p_2 \colon \mathbb{R} \to \mathbb{R}$ such that the number of rational points in $[0, 1]^k$ whose denominator is $\le d$ is $\le p_2(|\theta|)^k$. Evaluating $\mathsf{f}_{\neg\theta}$ at any such point takes a number of Turing steps $< p_3(|\theta|)$, for some polynomial p_3 only depending on k.

Now $T_{\text{check proper ideal}}$ carries out an exhaustive search over the polynomially few (in fact, fewer than d^{2k}) points in $[0, 1]^k$ of denominator $\le d$. This search terminates deciding problem (iii), after a number of steps smaller than $p_4(|\theta|)$, for a polynomial p_4 only depending on k. □

Assuming that the machine $T_{\text{check proper ideal}}$ has found some point in the zeroset of $\mathsf{f}_{\neg\theta}$, the decision procedure for the consistency of the conditional PSAT$_\infty$ problem may continue.

Proposition 9.15 *There is a Turing machine $T_{\text{polyhedron to polytope}}$ performing the following task: Having in input a finite list $\mathcal{P}$ of n-variable linear polynomials with integer coefficients, $T_{\text{polyhedron to polytope}}$ outputs the list of vertices of the bounded rational convex polyhedron $P = \{x \in [0, 1]^n \mid p(x) \ge 0 \text{ for all } p \in \mathcal{P}\}$.*

Proof Use Theorem 8.5 and Corollary 8.6. Observe the constructive character of the proofs. □

Using $T_{\text{polyhedron to polytope}}$ we proceed as follows:

Proposition 9.16 *There is a Turing machine $T_{\text{Mod-linearizer}}$ that performs the following task: On any input $\{(\psi_1, r_1), \ldots, (\psi_m, r_m), \theta\}$ of the conditional PSAT$_\infty$ problem (with the satisfiability of θ checked by $T_{\text{check proper ideal}}$), $T_{\text{Mod-linearizer}}$ outputs a rational triangulation ∇ of $[0, 1]^k$ with rational vertices $x_1, \ldots, x_w$ such that each function $\mathsf{f}_{\psi_1}, \ldots, \mathsf{f}_{\psi_m}, \mathsf{f}_\theta$ is linear on every simplex of ∇, and the subset*

$$\nabla_\theta = \{T \in \nabla \mid T \subseteq \mathsf{Mod}\,\theta\}$$

is a triangulation of $\mathsf{Mod}\,\theta$.

Proof A routine variant of the machine $T_{\text{linearizer}}$ of Lemma 8.7. □

On input $\{(\psi_1, r_1), \ldots, (\psi_m, r_m), \theta\}$ the Turing machine $T_{\text{Mod-linearizer}}$ has just constructed a triangulation ∇ of $[0, 1]^k$ with rational vertices $x_1, \ldots, x_w$ that linearizes every input function $\mathsf{f}_{\psi_1}, \ldots, \mathsf{f}_{\psi_m}, \mathsf{f}_\theta$. with $\nabla_\theta \subseteq \nabla$ a triangulation of $\mathsf{Mod}\,\theta$. The McNaughton function f_θ takes the constant value 1 precisely over its support $|\nabla_\theta|$ Every function $\mathsf{f}_{\psi_1}, \ldots, \mathsf{f}_{\psi_m} \in \mathcal{M}([0, 1]^k)$ is linear on each simplex of ∇_θ. Let

$$\{y_1, \ldots, y_v\} \subseteq \{x_1, \ldots, x_w\}$$

be the set of vertices of $\nabla_\theta \subseteq \nabla$. Let

$$\mathsf{eval}\colon y \in \mathsf{Mod}\,\theta \mapsto \eta_y \in \mathsf{hom}(\mathcal{M}([0,1]^k) \upharpoonright \mathsf{Mod}\,\theta)$$

be the bijection in Proposition 7.7.

The following generalization of Lemma 8.9 provides the final key step to prove Theorem 9.2.

Lemma 9.17 (∇_θ-inconsistency implies inconsistency.) *With the above notation, let us assume that there exists $s = (s_1, \ldots, s_m) \in \mathbb{R}^m$ satisfying the following condition:*

$$(\nabla_\theta\text{-}inconsistency)\text{: For each } t = 1, \ldots, v, \quad \sum_{j=1}^{m} s_j(r_j - \eta_{y_t}(\mathsf{f}_{\psi_j} \upharpoonright \mathsf{Mod}\,\theta)) < 0.$$

Then for each $\eta \in \mathsf{hom}(\mathcal{M}([0,1]^k) \upharpoonright \mathsf{Mod}\,\theta)$, $\sum_{j=1}^m s_j(r_j - \eta(\mathsf{f}_{\psi_j} \upharpoonright \mathsf{Mod}\,\theta)) < 0$.

Proof The proof can be extracted from the proof of Lemma 8.9 with very little extra effort. □

By Lemma 9.17, the search for a witness of inconsistency over the infinite set of homomorphisms $\mathsf{hom}(\mathcal{M}([0,1]^k) \upharpoonright \mathsf{Mod}\,\theta)$ can be equivalently restricted to the set $y_1, \ldots, y_v$ of vertices of ∇_θ.

Conclusion of the proof of Theorem 9.2. Conditions (1)–(2) in Remark 8.3 concerning grammatical correctness of the input are immediately verified, precisely as in the unconditional case. Condition (3) about the input set of pairs forming a function (so that it can be extended to a state) can be effectively checked, by letting the machine $T_{\mathsf{logically-equivalent-bis}}$ check that equal restrictions to $\mathsf{Mod}\,\theta$ of the functions f_j, $j = 1, \ldots, m$ are assigned equal betting odds. The Turing-computability of Condition (4) is verified by a routine variant of the proof of Lemma 8.8.

Once the preliminary conditions (1)–(4) in Remark 8.3 have been checked positively, to decide the conditional PSAT$_\infty$ problem in the formulation given by Corollary 9.13(vi), we resort to Lemma 9.17. Its effect is that inconsistency need only be checked on the finite set of homomorphisms corresponding to the vertices $y_1, \ldots, y_v$ of the triangulation ∇_θ of $\mathsf{Mod}\,\theta$ that linearizes the functions coded by the input formulas $\psi_1, \ldots, \psi_m$. By construction, f_θ takes the constant value 1 precisely on the rational polyhedron $\mathsf{Mod}\,\theta = |\nabla_\theta|$. The triangulation ∇_θ has been computed by $T_{\mathsf{Mod-linearizer}}$ on input $\psi_1, \ldots, \psi_m, \theta$.

As in (8.7), with v in place of w, we next compute the *rational* matrix

$$M_{tj} = r_j - \eta_{x_t}(\mathsf{f}_{\psi_j}) = r_j - \mathsf{f}_{\psi_j}(x_t), \quad t = 1, \ldots, v; \quad j = 1, \ldots, m.$$

We must now solve the system

$$Ms \quad = \underbrace{(z_1\,, \quad z_2\,, \quad \ldots \quad , z_v)}_{\text{vector whose coordinates are all } <0}.$$

of linear inequalities $Ms < 0$ with rational coefficients, in the unknowns $s = (s_1, \dots, s_m) \in \mathbb{R}^m$, precisely as in the final part of the proof of Theorem 8.2. ∇_θ-inconsistency amounts to the existence of a solution. By Lemma 9.17, the existence of a solution $s \in \mathbb{R}^m$ is also equivalent to the inconsistency of the *function* $\beta = \{(\mathsf{f}_{\psi_1} \upharpoonright \mathsf{Mod}\,\theta,\ r_1), \dots, (\mathsf{f}_{\psi_m} \upharpoonright \mathsf{Mod}\,\theta,\ r_m)\}$ in the MV-algebra $\mathcal{M}([0,1]^k) \upharpoonright \mathsf{Mod}\,\theta$. By Theorem 6.3 and Farkas' Lemma 2.6, this is in turn equivalent to the non-existence of a state of $\mathcal{M}([0,1]^k) \upharpoonright \mathsf{Mod}\,\theta$ extending β. Summing up, the non-existence of a solution s for the above system of inequalities $Ms < 0$ is equivalent to the existence of a state σ extending β which is a convex combination of the finite set of homomorphisms $\eta_{y_1}, \dots, \eta_{y_v}$.

The existence/non-existence of a solution for our system of rational inequalities can be decided using Linear Programming or Fourier-Motzkin elimination, as in the final part of the proof of Theorem 8.2.

The proof of Theorem 9.2 is now complete. □

Recall that any convex combination of homomorphisms of an MV-algebra A is a special case of a state of A. As we have seen, the state space of A coincides with the closure of the set of convex combinations of homomorphisms. We then have:

Corollary 9.18 *The conditional PSAT$_\infty$ problem is equivalent to the following problem: Does the Lindenbaum MV-algebra $\mathsf{Form}_k/\equiv_\theta$ admit a convex combination σ of homomorphisms of $\mathsf{Form}_k/\equiv_\theta$ into $[0,1]$ such that $r_j = \sigma(\psi_j/{\equiv_\theta})$ for each $j = 1, \dots, m$? By Theorem 8.2, also this problem is decidable.* □

The conclusion of the above proof of Theorem 9.2, in combination with Proposition 9.11 yields now the following result:

Corollary 9.19 *(i) For any finitely presented MV-algebra A and elements $h_1, \dots, h_m \in A$, a map $\beta\colon \{h_1, \dots, h_m\} \to [0,1] \cap \mathbb{Q}$ is consistent if and only if it is extendable to a* convex combination *of homomorphisms of A into* $[0,1] \cap \mathbb{Q}$.

(ii) For finitely presented MV-algebras the conditional consistency problem is uniformly decidable, *in the sense that a single Turing machine can decide the problem on any input $\{(\psi_1, r_1), \dots, (\psi_m, r_m);\ \theta\}$ of k-variable formulas $\psi_1, \dots, \psi_m, \theta$, for arbitrary integers $m, k > 0$ and* rational *numbers $r_1, \dots, r_m \in [0,1]$.* □

Reverting to the special case of MV-algebras given by boolean algebras we have:

Corollary 9.20 *(i) Let A be a finitely presented* boolean *algebra. (This amounts to saying that A is finite.) Let $h_1, \dots, h_m$ be elements of A. Then a map $\beta\colon \{h_1, \dots, h_m\} \to [0,1] \cap \mathbb{Q}$ is consistent if and only if it is extendable to a* convex combination *of homomorphisms of A into the two-element boolean algebra $\{0,1\}$.*

(ii) For finite boolean algebras the conditional consistency problem is uniformly decidable, *in the sense that a single Turing machine can decide the problem on any input $\{(\psi_1, r_1), \dots, (\psi_m, r_m);\ \theta\}$ of k-variable boolean formulas $\psi_1, \dots, \psi_m, \theta$, for arbitrary integers $m, k > 0$ and rational numbers $r_1, \dots, r_m \in [0,1]$.* □

Proof (i) As is well known, boolean algebras constitute a full subcategory of the category of MV-algebras. Let A be a finitely presented boolean algebra A, *qua* finitely presented MV-algebra. Let k be the number of variables of the $Ł_\infty$-formula θ presenting A. First of all, using Proposition 9.16, it is decidable whether $\mathsf{Mod}\,\theta$ is a nonempty set of vertices of the k-cube $[0, 1]^k$. This is a necessary and sufficient condition for θ to present a boolean algebra. Assuming this to be the case, by Proposition 9.11, we may visualize A as the restriction of the McNaughton functions of the free MV-algebra $\mathcal{M}([0, 1]^k)$ to a subset $S = S_A = \mathsf{Mod}\,\theta$ of the set $\{0, 1\}^k$ of vertices of the k-cube $[0, 1]^k$. This is so because the McNaughton functions of $\mathcal{M}([0, 1]^k)$ take their values in $\{0, 1\}$ precisely at any vertex of $[0, 1]^k$. Necessarily, A is finite. Letting a be the number of atoms of A it follows that S has precisely a elements, for some $a \leq 2^k$. Now A may be identified with the 2^a many boolean functions over S. By Proposition 1.5, A can be also realized as the set $\{0, 1\}^a$ of vertices of the cube $[0, 1]^a$. The homomorphisms of A into the two-element boolean algebra $\{0, 1\}$ correspond to the atoms of A as well as to the standard basis vectors in a-space $\mathbb{R}^a$. The geometric representation of the states of A given by Corollary 1.11 as convex combinations of these standard basis vectors concludes the proof of (i).

(ii) Using the geometric representation of A in the proof of (i), the uniform decidability result (ii) routinely follows from its MV-algebraic counterpart given by Corollary 9.19(ii). □

Remark 9.21 As remarked in the next section, for presentations of boolean algebras by boolean formulas, *i.e., presentations within the equational class of boolean algebras*, the literature offers stronger results than Corollary 9.20(ii), dealing with the case when the boolean algebra A is presented by a $Ł_\infty$-formula.

Historical and bibliographical notes. For consequence relations and Lindenbaum algebras in Łukasiewicz logic see TheBook, Sect. 4.

Proposition 9.9 is due to Hay [13] and Wójcicki [28].

In TheBook, 4.5.3, 4.6.10 one can find a proof of the following result: *For any k-variable formulas ϕ and θ, the consequence problem $\theta \vdash \phi$ (equivalently, $\theta \models \phi$) is decidable. In particular, the tautology problem is decidable, and so is the satisfiability/logical consistency problem for formulas.*

Proposition 9.14 has a stronger form stating that the satisfiability problem for Łukasiewicz logic is NP-complete. See [18].

A natural non-tarskian notion $\models_\partial$ of semantic consequence always coinciding with $\vdash$ in Łukasiewicz logic can be found in [20]. Intuitively, for an arbitrary set Θ of premises $\Theta \models_\partial \phi$ holds iff every valuation V satisfying all $\theta \in \Theta$ also satisfies ϕ (i.e., $\Theta \models \phi$) *and* whenever an infinitesimal perturbation $V + \mathrm{d}V$ satisfies all $\theta \in \Theta$ then $V + \mathrm{d}V$ also satisfies ϕ.

With reference to the infinite set of premises $\Theta = \{\theta_1, \theta_2, \dots\}$ in the proof of Theorem 9.5(ii), the condition $\{\theta_1, \theta_2, \dots\} \models_\partial \neg X$ fails because: On the one hand, $\frac{\partial \mathsf{f}_{\theta_n}}{\partial x}(0) = 0$, meaning that the satisfiability of f_{θ_n} is preserved under infinitesimal perturbations $0 \mapsto 0 + \mathrm{d}x$ on the right of the point 0. Each f_{θ_n} has value 1 at 0 and also over a suitably small open interval I_n containing 0. On the other hand,

$\frac{\partial f_{\neg X}}{\partial x}(0) = -1$, showing that the truth-value 1 of $f_{\neg X}$ at 0 is not preserved under infinitesimal perturbations $0 \mapsto 0 + dx$. As a particular case of the identity $\models_\partial\, = \,\vdash$ proved in [20], observe that $\{\theta_1, \theta_2, \dots\} \vdash \neg X$ also fails.

The piecewise linearity of every McNaughton function f ensures that one can compute the directional derivative of f along any direction. In the above example, taking the quotient of f by the maximal ideal $\mathfrak{m}$ given by all McNaughton functions vanishing at 0, amounts to evaluating f at 0. Taking the quotient by the (only) prime ideal $\mathfrak{p} \subsetneqq \mathfrak{m}$ amounts to also computing the directional derivative of f at 0 along the positive direction of the x-axis. For more information on the classification of prime ideals of $\mathcal{M}([0, 1]^k)$ in terms of directional derivatives and their refinements see [4] and [20].

The paper [23] presents an algorithmic approach to representing rational McNaughton functions in plain Łukasiewicz logic via representations modulo satisfiability. Mod θ.

For Minkowski convex body theorem in the geometry of numbers see [3] or [15]. For polyhedral topology also see [14]. The decidability of the conditional consistency problem was first proved in [19] using the Tarski-Seidenberg decision method for elementary algebra, [26, 27]. Also see [2] for a refinement of this result. For further developments to other logics see, e.g., [1, 5, 6, 8, 9, 22, 24].

In the approach to conditionals adopted in this chapter, the antecedent conditioning formula θ is assumed to be 100% true. Its effect is to restrict the set of "possible worlds" to those where the truth-value of θ is equal to 1. Other approaches take into account the fact that also the antecedent of a conditional may be [0, 1]-valued. See, e.g., [10, 16, 17].

With reference to Corollary 9.20, conditional probability is a central topic in *classical* probability theory. The paper [11] is a basic reference for PSAT and conditional PSAT. In the paper [7] one can find a proof that the conditional $PSAT_2$ problem is NP complete. In the paper [12] the satisfiability problem for fuzzy probability logic over Łukasiewicz logic is shown to be NP-complete. The same problem in the logic joining Łukasiewicz and product logic it is shown to be in PSPACE.

References

1. Aguzzoli, S., Gerla, B., Marra, V. (2008). De Finetti's no-Dutch-book criterion for Gödel logic. Special issue on "Many-valued Logic and Cognition". *Studia Logica, 90*, 25–41.
2. Bova, S., & Flaminio, T. (2010). The coherence of Łukasiewicz assessments is NP-complete. *International Journal of Approximate Reasoning, 51*, 294–304.
3. Brønsted, A. (1983). *An introduction to convex polytopes, Graduate Texts in Mathematics* (Vol. 90). Springer.
4. Busaniche, M., & Mundici, D. (2007). Geometry of Robinson consistency in Łukasiewicz logic. *Annals of Pure and Applied Logic, 147*, 1–22.
5. Fedel, M., Keimel, K., Montagna, F., & Roth, W. (2013). Imprecise probabilities, bets and functional analytic methods in Łukasiewicz logic. *Forum Mathematicum, 25*(2), 405–441.
6. Finger, M., & Preto, S. (2020). Probably partially true: Satisfiability for Łukasiewicz infinitely-valued probabilistic logic and related topics. *Journal of Automated Reasoning, 64*, 1269–1286.

7. Flaminio, T. (2007). NP-containment for the coherence test of assessments of conditional probability: a fuzzy logic approach. *Archive for Mathematical Logic, 46*, 301–319.
8. Flaminio, T., Godo, L., & Hosni, H. (2015). Coherence in the aggregate: A betting method for belief functions on many-valued events. *International Journal of Approximate Reasoning, 58*, 71–86.
9. Flaminio, T., Montagna, F. (2007). An algebraic approach to states on MV-algebras. In M. Štěpnička, et al. (Eds.), *Proceedings 5th EUSFLAT07 Conference* (Vol. 2, pp. 201–206), Ostrava (Czech Republic).
10. Gerla, B. (2000). Conditioning a state by a Łukasiewicz event: A probabilistic approach to Ulam games. *Theoretical Computer Science, 230*, 149–166.
11. Georgakopoulos, G., Kavvadias, D., & Papadimitriou, C. H. (1988). Probabilistic satisfiability. *Journal of Complexity, 4*, 1–11.
12. Hàjek, P., & Tulipani, S. (2001). Complexity of fuzzy probability logics. *Fundamenta Informaticae, 45*(3), 207–213.
13. Hay, L. S. (1963). Axiomatization of the infinite-valued predicate calculus. *Journal of Symbolic Logic, 28*, 77–86.
14. Hudson, J. F. P. (1969). *Piecewise linear topology*. W.A. Benjamin Inc.
15. Lekkerkerker, C. G. (1969). *Geometry of numbers*. North-Holland.
16. Montagna, F. (2012). Partially undetermined many-valued events and their conditional probability. *Journal of Philosophical Logic, 41*, 563–593.
17. Montagna, F., Fedel, M., & Scianna, G. (2013). Non-standard probability, coherence and conditional probability on many-valued events. *International Journal of Approximate Reasoning, 54*(5), 573–589.
18. Mundici, D. (1987). Satisfiability in many-valued sentential logic is NP-complete. *Theoretical Computer Science, 52*, 145–153.
19. Mundici, D. (2006). Bookmaking over infinite-valued events. *International Journal of Approximate Reasoning, 43*, 223–240.
20. Mundici, D. (2015). The differential semantics of Łukasiewicz syntactic consequence. In F. Montagna et al. (Eds.), *Petr Hájek on mathematical fuzzy logic. Outstanding contributions to logic* (Vol. 6, pp. 143–157). Springer.
21. Pap, E. (Ed.). (2002). *Handbook of Measure Theory, I*. North-Holland.
22. Preto, S., & Finger, M. (2023). Proving properties of binary classification neural networks via Łukasiewicz logic. *Logic Journal of the IGPL, 31*(5), 805–821.
23. Preto, S., & Finger, M. (2022). Efficient representation of piecewise linear functions into Łukasiewicz logic modulo satisfiability. *Math. Structures in Computer Science, 32*(9), 1119–1144.
24. Riečan, B., Mundici, D., *Probability on MV-algebras* (Vol. II, pp. 869–909). In [21].
25. Schrijver, A. (1986). *Theory of linear and integer programming*. Wiley.
26. Seidenberg, A. (1954). A new decision method for elementary algebra. *Annals of Mathematics, 60*, 365–374.
27. Tarski, A. (1951). *A decision method for elementary algebra and geometry*. University of California Press.
28. Wójcicki, R. (1977). On matrix representations of consequence operations of Łukasiewicz sentential calculi, Zeitschrift für math. Logik und Grundlagen der Mathematik, 19 (1973) 239–247. Reprinted. In R. Wójcicki & G. Malinowski (Eds.), *Selected Papers on Łukasiewicz Sentential Calculi* (pp. 101–111).

Chapter 10
Product and Exchangeable States

In the first part of the book our self-contained proof of de Finetti's characterization of probabilistic *consistency* has been applied to obtain the *product* law for the probability of conjunctions of algebraically *independent* boolean events. To record the possible outcomes of n coin tosses, in Chapter 1 §1.4 we introduced conjunctions of the form $X_1^{\mathsf{b}_1} \wedge \cdots \wedge X_n^{\mathsf{b}_n}$. Fix $p \in [0, 1]$. Let $\beta = \beta_p$ be the book in the free boolean algebra $\mathfrak{B}_n$ assigning to (the equivalence class t of the formula) $X_1^{\mathsf{b}_1} \wedge \cdots \wedge X_n^{\mathsf{b}_n}$ the *product* $p_1 \dots p_n$, where $p_i = p$ or $p_i = 1 - p$ according as $\mathsf{b}_i = 1$ or $\mathsf{b}_i = 0$. As befits a book of conjunctions of *independent* events in the free *product* $\mathfrak{B}_n$ of n *independent* copies of the free one-generated boolean algebra $\mathfrak{B}_1$, the map β is *consistent*. This follows from the boolean algebraic *product* law theorem. By the *consistency* theorem, β is extendable to a state of $\mathfrak{B}_n$. Indeed, β is extendable to a unique state, known as the *product* state of type p in $\mathfrak{B}_n$. The free boolean algebra $\mathfrak{B}_\omega$ inherits from its subalgebras $\mathfrak{B}_n$ a uniquely determined state, called the *product* state of $\mathfrak{B}_\omega$ of type p. Its value on each equivalence class $t \in \mathfrak{B}_n$ only depends on the number of bits b_i equal to 1 in $X_1^{\mathsf{b}_1} \wedge \cdots \wedge X_n^{\mathsf{b}_n}$. This property is known as *exchangeability*. We will characterize the set of exchangeable states of $\mathfrak{B}_\omega$ and obtain as a corollary a self-contained proof of de Finetti's celebrated exchangeability/representation theorem.

10.1 Product and Exchangeable States. Statement of the Main Theorem

When we flip a coin there is no logical dependence between the ith and the jth toss ($i \neq j$). Letting the symbols $X_1, X_2, \dots$ denote boolean variables, for any $i = 1, 2, \dots$ we write X_i or X_i^1 if the ith coin toss comes out "Head". Otherwise we write $\neg X_i$ or X_i^0. The free boolean algebra $\mathfrak{B}_\omega$ on a countably infinite set of generators is naturally identified with the boolean algebra on the free generating set given by the (always logical) equivalence classes of the variables X_i, $(i = 1, 2, \dots)$. The

D. Mundici, *Algebraic Probabilistic Consistency*, Trends in Logic 69,
https://doi.org/10.1007/978-3-031-98337-5_10

equivalence classes of the first n variables constitute a free generating set of $\mathfrak{B}_n$, and induce a natural embedding of $\mathfrak{B}_n$ into $\mathfrak{B}_{n+1}$ and into $\mathfrak{B}_\omega$. Thus without loss of generality we may write

$$\mathfrak{B}_1 \subseteq \mathfrak{B}_2 \subseteq \cdots \subseteq \mathfrak{B}_\omega = \bigcup \mathfrak{B}_n. \tag{10.1}$$

Using the above notation, for any $n = 1, 2, \ldots$ and $\mathsf{b}_i \in \{0, 1\}$, let the n-variable formula ϕ be given by

$$\phi = X_1^{\mathsf{b}_1} \wedge X_2^{\mathsf{b}_2} \wedge \cdots \wedge X_n^{\mathsf{b}_n}.$$

The equivalence class $t = t_\phi$ of ϕ is said to be a *minterm* of $\mathfrak{B}_n$, and ϕ is said to *code* t. The formula ϕ is uniquely determined by the minterm t, $\phi = \phi_t$. For any minterm t of $\mathfrak{B}_n$ we write $\mathsf{pos}(t)$ for the number of non-negated literals occurring in ϕ_t, and let $\mathsf{neg}(t) = n - \mathsf{pos}(t)$ be the number of negated literals. Trivially,

$$\text{any two distinct minterms } t', t'' \text{ of } \mathfrak{B}_n \text{ are } \textit{incompatible} : t' \wedge t'' = 0. \tag{10.2}$$

Letting $\mathsf{minterm}(\mathfrak{B}_n)$ denote the set of minterms of $\mathfrak{B}_n$, for all $m > n$ we have :

$$\mathsf{minterm}(\mathfrak{B}_n) \cap \mathsf{minterm}(\mathfrak{B}_m) = \emptyset \ \text{ and } \mathsf{minterm}(\mathfrak{B}_n) \subseteq \mathfrak{B}_m. \tag{10.3}$$

For every element $a \in \mathfrak{B}_\omega$, letting n_a be the smallest n such that a belongs to $\mathfrak{B}_n$ we can write

$$a = \bigvee \{t \in \mathsf{minterm}(\mathfrak{B}_{n_a}) \mid a \geq t\} = \bigvee \{u \in \mathsf{minterm}(\mathfrak{B}_m) \mid a \geq u\}, \tag{10.4}$$

for all $m \geq n_a$. Therefore, by (10.1), we may unambiguously define

$$\mathsf{minterm}(\mathfrak{B}_\omega) = \mathsf{minterm}(\mathfrak{B}_1) \cup \mathsf{minterm}(\mathfrak{B}_2) \cup \cdots .$$

By (10.3), for any $t \in \mathsf{minterm}(\mathfrak{B}_\omega)$ there is a uniquely determined integer $n = n_t$ such that t is a minterm of $\mathfrak{B}_n$.

The special element $\mathsf{a}_{n,k}$ *and the special state* $\xi_{n,k}$. For any $n = 1, 2, \ldots$ and $k = 0, 1, \ldots, n$, the *special element* $\mathsf{a}_{n,k}$ of $\mathfrak{B}_n$ is defined by

$$\mathsf{a}_{n,k} = \bigvee \{t \in \mathsf{minterm}(\mathfrak{B}_n) \mid \mathsf{pos}(t) = k\}. \tag{10.5}$$

Thus an element $r \in \mathsf{minterm}(\mathfrak{B}_n)$ satisfies $r \leq \mathsf{a}_{n,k}$ if and only if $\mathsf{pos}(r) = k$. In other words, the set of minterms of $\mathfrak{B}_n$ *majorized* by $\mathsf{a}_{n,k}$ is the $\binom{n}{k}$-element set of minterms of $\mathfrak{B}_n$ having precisely k positive literals. We next introduce the *special state* $\xi_{n,k}$ of $\mathfrak{B}_n$ by the stipulation:

$$\xi_{n,k}(t) = \frac{1}{\binom{n}{k}} \quad \text{for each } t \in \mathsf{minterm}(\mathfrak{B}_n) \text{ such that } t \leq \mathsf{a}_{n,k}. \tag{10.6}$$

By (10.2), the additivity and normalization properties of the state $\xi_{n,k}$ ensure that $\xi_{n,k}$ assigns the value 0 to the $2^n - \binom{n}{k}$ minterms r of $\mathfrak{B}_n$ with $\mathsf{pos}(r) \neq k$.

Product and exchangeable states.

Definition 10.1 For any $p \in [0, 1]$ the *product state* σ_p of $\mathfrak{B}_n$ *(of type p)* is the only state σ of $\mathfrak{B}_n$ which satisfies the identity

$$\sigma(t) = p^{\mathsf{pos}(t)}(1-p)^{\mathsf{neg}(t)}, \tag{10.7}$$

for any minterm t of $\mathfrak{B}_n$. For completeness, when both p and $\mathsf{pos}(t)$ are equal to 0 we set $p^{\mathsf{pos}(t)} = 1$. Symmetrically, when both $1-p$ and $\mathsf{neg}(t)$ are equal to 0 we let $(1-p)^{\mathsf{neg}(t)} = 1$. We say that a function $\sigma : \mathfrak{B}_\omega \to [0, 1]$ is a *product state of* $\mathfrak{B}_\omega$ *(of type p)* if for any $n = 1, 2, \ldots$, the restriction $\sigma \restriction \mathfrak{B}_n$ is a *product state of* $\mathfrak{B}_n$ *(of type p)*.

For any $p \in [0, 1]$ and $n = 1, 2, \ldots$, the free boolean algebra $\mathfrak{B}_n$ has precisely one product state of type p. Uniqueness follows from (10.4), because every boolean formula $\psi = \psi(X_1, \ldots, X_n)$ is equivalent to the sup of (pairwise incompatible, by (10.2)) minterms of $\mathfrak{B}_n$. Let $\sigma_{n,p}$ denote this state. For any $l \leq n$ the restriction to $\mathfrak{B}_l$ of $\sigma_{n,p}$ is the unique product state of $\mathfrak{B}_l$ of type p, in symbols,

$$\sigma_{l,p} = \sigma_{n,p} \restriction \mathfrak{B}_l.$$

The union of the sequence of states $\sigma_{1,p} \subseteq \sigma_{2,p} \subseteq \ldots$ is the unique product state of $\mathfrak{B}_\omega$ of type p. For any two minterms t, t' of $\mathfrak{B}_n$ the product state $\sigma_{n,p}$ gives the same value to t and t' if and only if t and t' have the same number of positive literals. The following definition provides a larger family of states of $\mathfrak{B}_n$ satisfying this latter property:

Definition 10.2 A state of σ of $\mathfrak{B}_\omega$ is said to be *exchangeable* if for every minterm t of $\mathfrak{B}_\omega$ the value $\sigma(t)$ only depends on the pair $(\mathsf{pos}(t), \mathsf{neg}(t))$. For every $n = 1, 2, \ldots$, the restriction $\sigma_n = \sigma \restriction \mathfrak{B}_n$ is said to be an *exchangeable state of* $\mathfrak{B}_n$.

As a first set of examples, from (10.2) and (10.4) it follows that the special state $\xi_{n,k}$ defined in (10.6) is an exchangeable state of $\mathfrak{B}_n$.

Statement of the boolean exchangeability theorem. Fix $n = 1, 2, \ldots$ and $k = 0, 1, \ldots, n$. Let ς be an exchangeable state of $\mathfrak{B}_n$. Then ς assigns the same value to each minterm t of $\mathfrak{B}_n$ majorized by the special element $\mathsf{a}_{n,k}$. Specifically,

$$\varsigma(t) = \frac{\varsigma(\mathsf{a}_{n,k})}{\binom{n}{k}}. \tag{10.8}$$

From the pairwise incompatibility (10.2) of the minterms of $\mathfrak{B}_n$ and the normalization and additivity properties of ς we then have:

$$\varsigma(\mathsf{a}_{n,0}) + \cdots + \varsigma(\mathsf{a}_{n,n}) = \varsigma(\mathsf{a}_{n,0} \vee \cdots \vee \mathsf{a}_{n,n}) = 1. \tag{10.9}$$

- In case $0 < \varsigma(\mathsf{a}_{n,k}) < 1$ we must have $\varsigma(\mathsf{a}_{n,k'}) > 0$ for at least one $k' \neq k$. By (10.6)–(10.9), ς is expressible as a nontrivial convex combination of the exchangeable states $\xi_{n,0}, \ldots, \xi_{n,n}$ of $\mathfrak{B}_n$, with coefficients equal to $\varsigma(\mathsf{a}_{n,0}), \ldots, \varsigma(\mathsf{a}_{n,n})$. (Convex combinations are taken in the finite-dimensional vector space $\mathbb{R}^{\mathfrak{B}_n}$.)
- On the other hand, in case $\varsigma(\mathsf{a}_{n,k}) = 1$ the state ς coincides with the special state $\xi_{n,k}$, and hence ς is not expressible as a nontrivial convex combination of exchangeable states of $\mathfrak{B}_n$.

Summing up:

Lemma 10.3 *For all $n = 1, 2, \ldots$ and $k = 0, 1, \ldots, n$ we have:*

(i) $\xi_{n,k}$ is the most general possible extreme state in the compact convex finite-dimensional subset of $\mathbb{R}^{\mathfrak{B}_n}$ consisting of all exchangeable states of $\mathfrak{B}_n$.

(ii) Any exchangeable state ς of $\mathfrak{B}_n$ is a convex combination of the special states $\xi_{n,0}, \ldots, \xi_{n,n}$ with coefficients $\varsigma(\mathsf{a}_{n,0}), \ldots, \varsigma(\mathsf{a}_{n,n})$. □

Any product state is exchangeable, and so is any convex combination of exchangeable states. More generally, from the compactness of the state space $\mathsf{S}(\mathfrak{B}_\omega) \subseteq [0, 1]^{\mathfrak{B}_\omega}$ we obtain:

Lemma 10.4 *Let us suppose that the state ς lies in the closure of the set of convex combinations of product states of $\mathfrak{B}_\omega$. Then ς is an exchangeable state of $\mathfrak{B}_\omega$.* □

The remainder of this chapter is devoted to proving the following nontrivial converse of Lemma 10.4:

Theorem 10.5 (Boolean exchangeability theorem.) *Any exchangeable state σ of $\mathfrak{B}_\omega$ lies in the closure of the set of convex combinations of product states of $\mathfrak{B}_\omega$.*

10.2 Proof of the Boolean Exchangeability Theorem. First Part

Notation. Until otherwise specified, n will henceforth denote an arbitrary integer ≥ 1, and N an arbitrary integer $> n$. Let σ_N be the restriction of σ to $\mathfrak{B}_N$,

$$\sigma_N = \sigma \restriction \mathfrak{B}_N.$$

Then σ_N is an exchangeable state of $\mathfrak{B}_N$. Lemma 10.3 combined with (10.5) and (10.9) yields a representation of σ_N as a convex combination of the special states $\xi_{N,K}$ defined in (10.6) with coefficients $\lambda_{N,K}$, $(K = 0, \ldots, N)$ satisfying

$$\lambda_{N,K} = \sigma_N(\mathsf{a}_{N,K}), \quad \sigma_N = \sum_{K=0}^{N} \lambda_{N,K} \cdot \xi_{N,K} \quad \text{and} \quad \sum_{K=0}^{N} \lambda_{N,K} = 1. \tag{10.10}$$

Since the restriction map $\tau \in \mathsf{S}(\mathfrak{B}_N) \mapsto \tau \restriction \mathfrak{B}_n \in \mathsf{S}(\mathfrak{B}_n)$ preserves convex combinations, letting σ_n be shorthand for $\sigma \restriction \mathfrak{B}_n$, we have

$$\sigma_n = (\sigma \restriction \mathfrak{B}_N) \restriction \mathfrak{B}_n = \sum_{K=0}^{N} \lambda_{N,K} \cdot (\xi_{N,K} \restriction \mathfrak{B}_n). \tag{10.11}$$

By (10.10)–(10.11), each $t \in \mathsf{minterm}(\mathfrak{B}_n)$ satisfies the identities

$$\sigma(t) \;=\; \sigma_N(t) \;=\; \sigma_n(t) \;=\; \sum_{K=0}^{N} \lambda_{N,K} \cdot \xi_{N,K}(t) = \sum_{K=0}^{N} \sigma(\mathsf{a}_{N,k}) \cdot \xi_{N,K}(t). \tag{10.12}$$

Let us assume that t is coded by the formula $X_1^{\mathsf{b}_1} \wedge \cdots \wedge X_n^{\mathsf{b}_n}$ for some n-tuple of bits $(\mathsf{b}_1, \ldots, \mathsf{b}_n) \in \{0, 1\}^n$. By (10.1) and (10.3), t is a member of $\mathfrak{B}_N$. Likewise, let $u \in \mathsf{minterm}(\mathfrak{B}_N)$ be coded by $X_1^{\mathsf{b}'_1} \wedge \cdots \wedge X_N^{\mathsf{b}'_N}$, for bits $\mathsf{b}'_1, \ldots, \mathsf{b}'_N$. We then have

$$t \geq u \text{ (read: ``} t \textit{ majorizes } u\text{'') iff } \mathsf{b}_1 = \mathsf{b}'_1, \ldots, \mathsf{b}_n = \mathsf{b}'_n.$$

As a consequence, if t majorizes u then

$$\mathsf{neg}\,(t) \leq \mathsf{neg}\,(u) \quad \text{and} \quad \mathsf{pos}\,(t) \leq \mathsf{pos}\,(u) \leq N - \mathsf{neg}\,(t). \tag{10.13}$$

A straightforward argument then shows that for each $K = \mathsf{pos}\,(t), \ldots, N - \mathsf{neg}\,(t)$, the number of minterms v of $\mathfrak{B}_N$ with $\mathsf{pos}\,(v) = K$ and $t \geq v$ is equal to $\binom{N-n}{K-\mathsf{pos}\,(t)}$. By (10.6), for any such v we have

$$\xi_{N,K}(v) = \frac{1}{\binom{N}{K}}.$$

By (10.3)-(10.4), the minterm $t \in \mathsf{minterm}(\mathfrak{B}_n) \subseteq \mathfrak{B}_N$ is the disjunction of all minterms of $\mathfrak{B}_N$ majorized by t. Hence,

$$\xi_{N,K}(t) = \begin{cases} \dfrac{\binom{N-n}{K-\mathsf{pos}\,(t)}}{\binom{N}{K}} & \text{if } K \in \{\mathsf{pos}\,(t), \ldots, N - \mathsf{neg}\,(t)\} \\ 0 & \text{if } K \in \{0, \ldots, \mathsf{pos}\,(t) - 1\} \cup \{N - \mathsf{neg}\,(t) + 1, \ldots, N\}. \end{cases}$$

For any $n = 1, 2, \ldots,$ minterm t of $\mathfrak{B}_n$ and $N > n$, from (10.12) we get

$$\sigma(t) \;=\; \sum_{K=\mathsf{pos}\,(t)}^{N-\mathsf{neg}\,(t)} \lambda_{N,K} \cdot \frac{\binom{N-n}{K-\mathsf{pos}\,(t)}}{\binom{N}{K}}, \tag{10.14}$$

with each coefficient $\lambda_{N,K} \in [0, 1]$ as in (10.10). By (10.3)–(10.4), any minterm t of $\mathfrak{B}_n \subseteq \mathfrak{B}_N$ coincides with the disjunction of all minterms of $\mathfrak{B}_N$ majorized by t. The convex combination $\sum_{K=0}^{N} \lambda_{N,K}\, \pi_{K/N}$ of the product states $\pi_{K/N}$ of $\mathfrak{B}_\omega$ is a state of $\mathfrak{B}_\omega$. Upon restricting $\sum_{K=0}^{N} \lambda_{N,K}\, \pi_{K/N}$ to $\mathfrak{B}_N$ (resp., to $\mathfrak{B}_n$) we obtain a state of $\mathfrak{B}_N$ (resp., a state of $\mathfrak{B}_n$).

10.3 Proof of the Boolean Exchangeability Theorem. Second Part

Claim: For every $n = 1, 2, \ldots$ and $\epsilon > 0$ there is an integer $N = N_{n,\epsilon} > n$ such that the convex combination $\gamma = \sum_{K=0}^{N} \lambda_{N,K}\, \pi_{K/N}$ agrees with σ over $\mathsf{minterm}(\mathfrak{B}_n)$ up to an error $< \epsilon$. Thus for every $t \in \mathsf{minterm}(\mathfrak{B}_n)$,

$$|\sigma(t) - \gamma(t)| = \left|\sigma(t) - \sum_{K=0}^{N} \lambda_{N,K}\, (K/N)^{\mathsf{pos}\,(t)}\, (1 - K/N)^{n - \mathsf{pos}\,(t)}\right| < \epsilon. \tag{10.15}$$

Proof of Claim. By way of contradiction, let $n \in \mathbb{Z}_{\geq 1}$ and $\epsilon \in \mathbb{R}_{>0}$ be counterexamples to our claim. Thus n is a fixed integer ≥ 1 and ϵ is a fixed real number > 0 such that for every $N = n+1, n+2, \ldots$ some minterm $t = t_N \in \mathfrak{B}_n$ satisfies the inequality

$$\left|\sigma(t) - \sum_{K=0}^{N} \lambda_{N,K}\, \pi_{K/N}(t)\right| \geq \epsilon. \tag{10.16}$$

Without loss of generality, $\epsilon < 1$.
Since the boolean algebra $\mathfrak{B}_n$ is finite, some *fixed* minterm $t^* \in \mathfrak{B}_n$ satisfies

$$\left|\sigma(t^*) - \sum_{K=0}^{N} \lambda_{N,K}\, \pi_{K/N}(t^*)\right| \geq \epsilon \;\text{ for infinitely many } N > n. \tag{10.17}$$

Case 1: $\mathsf{pos}\,(t^*) \notin \{0, n\}$.

Let k be shorthand for $\mathsf{pos}\,(t^*)$. Then $k \in \{1, \ldots, n-1\}$. By (10.13) and (10.14), for infinitely many $N > n$ have the identities

$$\sigma(t^*) = \sum_{K=k}^{N-n+k} \lambda_{N,K} \frac{\binom{N-n}{K-k}}{\binom{N}{K}}$$

$$= \sum_{K=k}^{N-n+k} \lambda_{N,K} \frac{(N-n)!}{(K-k)!(N-n-(K-k))!} \cdot \frac{K!(N-K)!}{N!}$$

$$= \sum_{K=k}^{N-n+k} \lambda_{N,K} \frac{(K-k+1)\cdots K}{(N-n+1)\cdots N} \cdot (N-n-(K-k)+1)\cdots(N-K)$$

$$= \sum_{K=k}^{N-n+k} \lambda_{N,K} \frac{(K-k+1)\cdots K}{(N-n+1)\cdots(N-n+k)} \cdot$$

$$\cdot \frac{(N-K-n+k+1)\cdots(N-K)}{(N-n+k+1)\cdots N}$$

$$= \sum_{K=k}^{N-n+k} \lambda_{N,K} \prod_{i=1}^{k} \frac{K-k+i}{N-n+i} \prod_{j=1}^{n-k} \frac{N-n-K+k+j}{N-n+k+j},$$

whence

$$\sigma(t^*) = \sum_{K=k}^{N-n+k} \lambda_{N,K} \prod_{i=1}^{k} \frac{K/N-(k-i)/N}{1-(n-i)/N} \prod_{j=1}^{n-k} \frac{1-K/N-(n-k-j)/N}{1-(n-k-j)/N}. \tag{10.18}$$

Since $0 \leq \lambda_{N,K} \leq 1$, our standing hypothesis $k \notin \{0, n\}$ yields

$$0 = \lim_{N\to\infty} \sum_{K=0}^{k-1} \lambda_{N,K} (K/N)^k (1-K/N)^{n-k}$$

$$= \lim_{N\to\infty} \sum_{K=N-n+k+1}^{N} \lambda_{N,K} (K/N)^k (1-K/N)^{n-k}.$$

Then by (10.17),

$$\epsilon \leq \left|\sigma(t^*) - \sum_{K=0}^{N} \lambda_{N,K}\, \pi_{K/N}(t^*)\right| \leq \left|\sigma(t^*) - \sum_{K=k}^{N-n+k} \lambda_{N,K}\, \pi_{K/N}(t^*)\right|$$

$$+ \sum_{K=0}^{k-1} \lambda_{N,K}\, \pi_{K/N}(t^*) \quad + \sum_{K=N-n+k+1}^{N} \lambda_{N,K}\, \pi_{K/N}(t^*)$$

$$\leq \left|\sigma(t^*) - \sum_{K=k}^{N-n+k} \lambda_{N,K}\, \pi_{K/N}(t^*)\right| + \epsilon/2.$$

By (10.18), for infinitely many $N > n$ we then have

$$\epsilon/2 \leq \Bigg| \sum_{K=k}^{N-n+k} \lambda_{N,K} \Bigg(\prod_{i=1}^{k} \frac{K/N - (k-i)/N}{1-(n-i)/N} \cdot \tag{10.19}$$

$$\cdot \prod_{j=1}^{n-k} \frac{1 - K/N - (n-k-j)/N}{1-(n-k-j)/N} - \pi_{K/N}(t^*)\Bigg)\Bigg|.$$

Next for any

$$N > n,\ \ i \in \{1, \ldots, k\},\ \ j \in \{1, \ldots, n-k\},\ \text{and}\ \ K \in \{k, \ldots, N-n+k\},$$

let the rational numbers $c_i = c_i(N, K)$ and $d_j = d_j(N, K)$ be defined by

$$c_i = (1 - \frac{k-i}{K})\Big/(1 - \frac{n-i}{N}) \text{ and } d_j = (1 - \frac{n-k-j}{N-K})\Big/(1 - \frac{n-k-j}{N}). \tag{10.20}$$

Then our standing hypothesis $k \in \{1, \ldots, n-1\}$ immediately yields

$$0 \leq 1 - \frac{k}{K} \leq c_i \leq \frac{1}{1-n/N} \quad \text{and} \quad 0 \leq 1 - \frac{n-k}{N-K} \leq d_j \leq 1. \tag{10.21}$$

By (10.20), for infinitely many $N > n$ we can rewrite (10.19) as follows:

$$\epsilon/2 \leq \sum_{K=k}^{N-n+k} \lambda_{N,K} \Bigg| \prod_{i=1}^{k} \frac{K/N - (k-i)/N}{1-(n-i)/N} \prod_{j=1}^{n-k} \frac{1 - K/N - (n-k-j)/N}{1-(n-k-j)/N} -$$

$$- \pi_{K/N}(t^*) \Bigg|$$

$$= \sum_{K=k}^{N-n+k} \lambda_{N,K} \left| \prod_{i=1}^{k} (c_i\, K/N) \prod_{j=1}^{n-k} (d_j\, (1 - K/N)) \; - \; (K/N)^k\, (1 - K/N)^{n-k} \right|$$

$$= \sum_{K=k}^{N-n+k} \lambda_{N,K} \left| (K/N)^k\, (1 - K/N)^{n-k} \prod_{i=1}^{k} c_i \prod_{j=1}^{n-k} d_j - (K/N)^k\, (1 - K/N)^{n-k} \right|$$

and hence,

$$\epsilon/2 \;\le\; \sum_{K=k}^{N-n+k} \lambda_{N,K}\, (K/N)^k\, (1 - K/N)^{n-k} \left| \prod_{i=1}^{k} c_i \prod_{j=1}^{n-k} d_j - 1 \right|. \tag{10.22}$$

By (10.21), for all $N > n$ we have the sandwich inequalities

$$\left(1 - \frac{k}{K}\right)^k \cdot \left(1 - \frac{n-k}{N-K}\right)^{n-k} \;\le\; \prod_{i=1}^{k} c_i \prod_{j=1}^{n-k} d_j \le \left(\frac{1}{1 - n/N}\right)^k \cdot 1^{n-k}. \tag{10.23}$$

Furthermore, for all $N \ge 2n$, from (10.21) it follows that $c_i \;\le\; 2$, whence

$$0 \;\le\; \prod_{i=1}^{k} c_i(N, K) \prod_{j=1}^{n-k} d_j(N, K) \;\le\; 2^k \text{ for every } K \in \{k, \ldots, N - n + k\}. \tag{10.24}$$

10.4 Proof of the Boolean Exchangeability Theorem. Third Part

Let now η be a fixed but otherwise arbitrary real number satisfying

$$0 < \eta < 1/3. \tag{10.25}$$

Then, for all sufficiently large N, the integers $n > 0$ and $0 < k < n$ satisfy

$$k < \lfloor \eta N \rfloor < \lceil N - \eta N \rceil - 1 < N - n + k, \tag{10.26}$$

with $\lfloor x \rfloor$, (resp., $\lceil x \rceil$) shorthand for the largest integer $\le x$ (resp., the smallest integer $\ge x$).

Let us say that an integer K is ηN*-special* if

$$\lfloor \eta N \rfloor + 1 \leq K \leq \lceil N - \eta N \rceil. \tag{10.27}$$

Then for any integer $N > n$ and ηN-special K, the sandwich inequalities (10.23) yield

$$\frac{1}{(1-n/N)^k} \geq \left| \prod_{i=1}^{k} c_i(N, K) \prod_{j=1}^{n-k} d_j(N, K) \right| \tag{10.28}$$

$$\geq \left(1 - \frac{k}{K}\right)^k \left(1 - \frac{n-k}{N-K}\right)^{n-k} \tag{10.29}$$

$$\geq \left(1 - \frac{k}{1 + \lfloor \eta N \rfloor}\right)^k \left(1 - \frac{n-k}{N - \lceil (1-\eta) N \rceil}\right)^{n-k}. \tag{10.30}$$

Since, for all integers $N > n$ and $K \in [k,\ N-n-k]$, the rationals $c_i(N, K)$ and $d_j(N, K)$ satisfy (10.21), then for any η satisfying (10.25) there is an integer $M = M(\eta) > n$ such that the inequalities in (10.26) hold for all integers $N \geq M$. Thus by (10.27)-(10.30), for every $\theta > 0$ there is an integer $D(\theta, \eta) \geq M$ such that, for any ηN-special K,

$$\left| 1 - \prod_{i=1}^{k} c_i(N, K) \prod_{j=1}^{n-k} d_j(N, K) \right| < \theta \quad \text{for every integer } N > D(\theta, \eta).$$

In particular, for $\theta = \eta$, the following holds: For every integer $N > D(\eta, \eta)$ and ηN-special integer K,

$$\left| 1 - \prod_{i=1}^{k} c_i(N, K) \prod_{j=1}^{n-k} d_j(N, K) \right| < \eta. \tag{10.31}$$

We are now in a position to conclude the proof of our claim. Since $\epsilon/2$ satisfies (10.22), and ϵ, n, k are fixed, for infinitely many $N > n/\eta$ we have

$$
\begin{aligned}
\epsilon/2 \ \leq\ & \sum_{K=k}^{\lfloor \eta N \rfloor} \lambda_{N,K} \cdot (K/N)^k \cdot (1-K/N)^{n-k} \cdot \left| 1 - \prod_{i=1}^{k} c_i \prod_{j=1}^{n-k} d_j \right| \\
&+ \sum_{K=\lfloor \eta N \rfloor+1}^{\lceil N-\eta N \rceil} \lambda_{N,K} \cdot (K/N)^k \cdot (1-K/N)^{n-k} \cdot \left| 1 - \prod_{i=1}^{k} c_i \prod_{j=1}^{n-k} d_j \right| \\
&+ \sum_{K=\lceil N-\eta N \rceil+1}^{N-n+k} \lambda_{N,K} \cdot (K/N)^k \cdot (1-K/N)^{n-k} \cdot \left| 1 - \prod_{i=1}^{k} c_i \prod_{j=1}^{n-k} d_j \right| \\
\leq\ & \sum_{K=k}^{\lfloor \eta N \rfloor} \lambda_{N,K} \cdot \left(\frac{\eta N}{N}\right)^k \cdot 1 \cdot (1+2^k)\,, \qquad \text{by (10.24)} \\
&+ \sum_{K=\lfloor \eta N \rfloor+1}^{\lceil N-\eta N \rceil} \lambda_{N,K} \cdot 1 \cdot 1 \cdot \eta,\ \text{by (10.31), since any } K \text{ is } \eta N\text{-special here} \\
&+ \sum_{K=\lceil N-\eta N \rceil+1}^{N-n+k} \lambda_{N,K} \cdot 1 \cdot \left(1 - \frac{N-\eta N}{N}\right)^{n-k} \cdot (1+2^k)\,,\ \text{by (10.24)} \\
\leq\ & \sum_{K=k}^{\lfloor \eta N \rfloor} \lambda_{N,K} \cdot \eta \cdot 1 \cdot 3^k, \qquad \text{since } 0<\eta<1/3 \ \text{ by (10.25)} \\
&+ \sum_{K=\lfloor \eta N \rfloor+1}^{\lceil N-\eta N \rceil} \lambda_{N,K} \cdot 1 \cdot 1 \cdot 3^k\,\eta \\
&+ \sum_{K=\lceil N-\eta N \rceil+1}^{N-n+k} \lambda_{N,K} \cdot 1 \cdot \eta \cdot 3^k,\ \text{by (10.25)}.
\end{aligned}
$$

Therefore, for every $0 < \eta < 1/3$ infinitely many integers $N > n/\eta$ satisfy the inequality

$$
\epsilon/2 \ \leq \eta \cdot 3^k \cdot \sum_{K=k}^{N-n+k} \lambda_{N,K}.
$$

Let us recall that each one of ϵ, n, k is fixed, with $0 < \epsilon < 1$ and $1 \leq k \leq n-1$. Furthermore, by (10.10), $\sum_{K=0}^{N} \lambda_{N,K} = 1$. Setting now $\eta = \epsilon/3^{k+1}$, there is (an infinite set of integers) $N > n/\eta$ such that

$$
\frac{\epsilon}{2} \leq \eta \cdot 3^k = \frac{3^k \epsilon}{3^{k+1}} = \frac{\epsilon}{3},
$$

which shows that Case 1 is impossible.

Case 2: $\mathsf{pos}(t^*) \in \{0, n\}$. Then a routine simplification of the proof of Case 1 shows that also this case is impossible.
Our claim is thus settled.

10.5 Conclusion of the Proof of the Exchangeability Theorem

For each state $\rho \in \mathcal{S}(\mathfrak{B}_\omega)$, real number $\epsilon > 0$, and finite set $A \subseteq \mathfrak{B}_\omega$, let

$$\mathcal{N}_{\rho,\epsilon,A} = \{\varsigma \in \mathcal{S}(\mathfrak{B}_\omega) \mid \ |\varsigma(a) - \rho(a)| < \epsilon \text{ for each } a \in A\}. \tag{10.32}$$

$\mathcal{N}_{\rho,\epsilon,A}$ is an open neighborhood of ρ in the compact Hausdorff subspace $\mathcal{S}(\mathfrak{B}_\omega)$ of the Hilbert cube $[0, 1]^{\mathfrak{B}_\omega}$. Furthermore, for every open neighborhood $\mathcal{O} \subseteq \mathcal{S}(\mathfrak{B}_\omega)$ of ρ there is an $\epsilon > 0$ and a finite subset A of $\mathfrak{B}_\omega$ such that $\mathcal{N}_{\rho,\epsilon,A} \subseteq \mathcal{O}$.

There remains to prove that every open neighborhood $\mathcal{N}_{\sigma,\epsilon,A}$ of the exchangeable state σ of $\mathfrak{B}_\omega$ contains a convex combination of product states of $\mathfrak{B}_\omega$.

To this purpose, for every $n = 1, 2, \ldots$ and $\epsilon > 0$ our claim yields an integer $N > n$ and a convex combination γ of product states of $\mathfrak{B}_N$ such that $|\sigma(t) - \gamma(t)| < \epsilon$ for every $t \in \mathsf{minterm}(\mathfrak{B}_n)$. The state γ is the restriction to $\mathfrak{B}_N$ of a convex combination of product states of $\mathfrak{B}_\omega$. Hence, recalling (10.32), there is a choice function

$$\phi\colon \mathbb{Z}_{\geq 1} \times \mathbb{R}_{>0} \to \text{ the set of convex combinations of product states of } \mathfrak{B}_\omega$$

with the following property: For all $n = 1, 2, \ldots$ and $\epsilon > 0$, $\phi(n, \epsilon)$ is a convex combination of product states of $\mathfrak{B}_\omega$ lying in $\mathcal{N}_{\sigma,\epsilon,\mathsf{minterm}(\mathfrak{B}_n)}$. Replacing ϵ by $\epsilon/2^n$ and letting the state α of $\mathfrak{B}_\omega$ be defined by $\alpha = \phi(n, \epsilon/2^n)$, it follows that

$$\alpha \in \mathcal{N}_{\sigma,\,\epsilon/2^n,\,\mathsf{minterm}(\mathfrak{B}_n)}.$$

Since both α and σ are additive on the 2^n element set of pairwise incompatible minterms of $\mathfrak{B}_n$, then automatically $\alpha \in \mathcal{N}_{\sigma,\epsilon,\mathfrak{B}_n}$. In synthesis,

$$|\phi(n, \epsilon/2^n) - \sigma| < \epsilon/2^n \text{ over } \mathsf{minterm}(\mathfrak{B}_n) \implies |\phi(n, \epsilon/2^n) - \sigma| < \epsilon \text{ over } \mathfrak{B}_n.$$

In particular, for any $A = \{a_1, \ldots, a_z\} \subseteq \mathfrak{B}_\omega$ let $n(A)$ be the smallest integer n such that $A \subseteq \mathfrak{B}_n$. For every $\epsilon > 0$ let the convex combination χ of product states of $\mathfrak{B}_\omega$ be defined by

$$\chi = \phi(n(A),\ \epsilon/2^{n(A)}).$$

Then χ is a member of $\mathcal{N}_{\sigma,\epsilon,A}$, and for each $i = 1, \ldots, z$ we may write $|\sigma(a_i) - \chi(a_i)| < \epsilon$. As we have already noted, for every open neighborhood $\mathcal{O} \subseteq \mathcal{S}(\mathfrak{B}_\omega)$ of σ there is $\epsilon > 0$ along with a finite subset A of $\mathfrak{B}_\omega$ such that $\mathcal{N}_{\sigma,\epsilon,A} \subseteq \mathcal{O}$.

We conclude that σ lies in the closure of the set of convex combinations of product states of $\mathfrak{B}_\omega$ in the vector space $\mathbb{R}^{\mathfrak{B}_\omega}$ equipped with the product topology. This completes the proof of Theorem 10.5. □

Corollary 10.6 *A state σ of the free boolean algebra $\mathfrak{B}_\omega$ is exchangeable if and only if every open neighborhood of σ in the vector space $\mathbb{R}^{\mathfrak{B}_\omega}$ equipped with the product topology contains a convex combination of product states of $\mathfrak{B}_\omega$.* □

Corollary 10.7 *Let $\mathcal{P}([0,1])$ denote the compact Hausdorff space of regular Borel probability measures on the unit interval $[0,1]$. Then for any exchangeable state σ of $\mathfrak{B}_\omega$ there is a unique $\mu = \mu_\sigma \in \mathcal{P}([0,1])$ such that*

$$\text{for every } t \in \mathsf{minterm}(\mathfrak{B}_\omega),\ \sigma(t) = \int_{[0,1]} p^{\mathsf{pos}(t)}(1-p)^{\mathsf{neg}(t)}\,\mathrm{d}\mu. \tag{10.33}$$

Proof For every $n, m \in \mathbb{Z}_{\geq 1}$ the proof of Theorem 10.5 yields an integer $N > n$, along with rationals $0 \leq \lambda_{N,0}, \ldots, \lambda_{N,N}$ summing up to 1, and a probability measure $\mu_{n,m} \in \mathcal{P}([0,1])$ such that for each $t \in \mathsf{minterm}(\mathfrak{B}_n)$,

$$\left|\sigma(t) - \sum_{K=0}^{N} \lambda_{N,K}(K/N)^{\mathsf{pos}(t)}(1-K/N)^{n-\mathsf{pos}(t)}\right|$$

$$= \left|\sigma(t) - \int_{[0,1]} p^{\mathsf{pos}(t)}(1-p)^{\mathsf{neg}(t)}\,\mathrm{d}\mu_{n,m}\right| < 1/m.$$

Since $\mathcal{P}([0,1])$ (equipped with the weak topology) is sequentially compact, there is a subsequence $(\mu_{n,l})$ of $(\mu_{n,m})$ converging to some $\mu_n \in \mathcal{P}([0,1])$ such that, for each $t \in \mathsf{minterm}(\mathfrak{B}_n)$ and l,

$$\left|\sigma(t) - \int_{[0,1]} p^{\mathsf{pos}(t)}(1-p)^{\mathsf{neg}(t)}\,\mathrm{d}\mu_{n,l}\right| < 1/l.$$

As a consequence, for each $t \in \mathsf{minterm}(\mathfrak{B}_n)$,

$$\sigma(t) = \int_{[0,1]} p^{\mathsf{pos}(t)}(1-p)^{\mathsf{neg}(t)}\,\mathrm{d}\mu_n.$$

Since $\mathfrak{B}_\omega = \bigcup \mathfrak{B}_n = \bigcup \mathfrak{B}_l$, another application of the sequential compactness of $\mathcal{P}([0,1])$ yields a subsequence (μ_q) of (μ_n) converging to a probability measure $\mu = \mu_\sigma \in \mathcal{P}([0,1])$ that satisfies (10.33). The *uniqueness* of μ_σ follows from the Weierstrass approximation theorem. □

Historical and bibliographical notes. The proofs of the main results in this chapter were first published by the author in his paper [13]. They are reproduced here under license by Springer Nature.

Readers familiar with de Finetti's 1930 paper [4], or with its antecedent [3], may regard Corollary 10.6 as a minimalist mutation of the celebrated "de Finetti theorem", dispensing with such notions as: Bernoulli random variable, event, expectation, conditional probability distribution, moment, martingale, joint mass function, mixture, latent variable, independent and identically distributed variables, integral over a measure, which are commonly found in the multifarious [1, p. 9 §2] versions of this theorem. (For a first taste, try e.g., [1, 3.1, p. 19], [2, Theorem 35.10], [6, Theorem 9.16], [8, §2], [9, Theorem 2], [10, Theorem 12.24].)

The elementary proof of Theorem 10.5 given here directly yields Corollary 10.7, which comes closest to the original formulation of de Finetti theorem. He proved the existence of μ_σ in the Appendix of [4, pp. 124–133] using characteristic functions, the tools available to him when he communicated this result in the 1928 International Congress of Mathematicians in Bologna. (See [3].) In his subsequent papers on exchangeability he abandoned this formalism, using distribution functions, moments, and laws of large numbers. In [4, Chapter 4, §31, p. 121], the elements $X_1, X_2, \dots \in F_\omega$ of our minterms t are said to constitute a "class of equivalent events".

In recent years, the mutual relations between logic and Finetti's exchangeability theorem has intensively studied by the logic group in Salerno (Italy). See, e.g., [5, 11, 12].

References

1. Aldous, D. J. (1985). Exchangeability and related topics. Lecture Notes in Mathematics (Vol. 1117, pp. 1–198). Springer, Berlin, Heidelberg. https://doi.org/10.1007/BFb0099421.
2. Billingsley, P. (1995). Probability and measure (3rd ed.). Wiley.
3. de Finetti, B. (2015). Funzione caratteristica di un fenomeno aleatorio, Atti del Congresso Internazionale dei Matematici, Bologna 3–10 September, (1928), Tomo VI, pp. 179–190. Originally published by Nicola Zanichelli Editore S. P. A., Bologna. English translation. In: D. Alvarez-Melis, T. Broderick, A translation of "The characteristic function of a random phenomenon" by Bruno de Finetti. arXiv:1512.01229v1.
4. de Finetti, B. (2006). Funzione caratteristica di un fenomeno aleatorio, Memorie della Reale Accademia Nazionale dei Lincei (Vol. IV, Fascicolo 5, pp. 86–133). Reprinted. In: de Finetti, B. (1930). Opere Scelte (Vol. 1, pp. 109–157). Unione Matematica Italiana, Edizioni Cremonese, Florence.
5. Di Nola, A., & Dvurečenskij, & A., Lapenta, S. (2021). An approach to stochastic processes via non-classical logic. Annals of Pure and Applied Logic, 172(9), Article 103012.
6. Kallenberg, O. (2005). *Probabilistic symmetries and invariance principles*. Springer.
7. Kelley, J. L. (1959). Measures on Boolean algebras. *Pacific Journal of Mathematics, 9*, 1165–1177.
8. Kingman, J. F. K. (1978). Uses of Exchangeability. *Annals of Probability, 6*(2), 183–197.
9. Kirsch, W. (2019). An elementary proof of de Finetti' s theorem. *Statistics and Probability Letters, 151*, 84–88.
10. Klenke, A. (2013). *Probability theory*. A comprehensive course: Second edition, Universitext. Springer, Berlin.
11. Lapenta, S. (2022). de Finetti's coherence and exchangeability in infinitary logic. *International Journal of Approximate Reasoning, 145*, 36–50.
12. Lapenta, S., & Lenzi, G. (2022). Models, coproducts and exchangeability: Notes on states on Baire functions. *Mathematica Slovaca, 72*(4), 847–868.
13. Mundici, D. (2024). Permutation invariant Boolean states, Algebra Universalis, 85, article number 27. https://doi.org/10.1007/s00012-024-00859-3.

Chapter 11
Problems

(1) Extend to MV-algebras the boolean exchangeability theorem.
(2) Without assuming continuity, prove that product is the only consistency preserving operation $\diamond : [0, 1]^2 \to [0, 1]$.
(3) On any input $I = \{(\psi_1, r_1), \ldots, (\psi_m, r_m)\}$ of the PSAT_∞ problem, the MaxPSAT_∞ problem asks what is the maximum number of elements of I forming a consistent book. Evaluate the complexity of MaxPSAT_∞.
(4) On any input $\{(\psi_1, r_1), \ldots, (\psi_m, r_m)\}$ of the PSAT_∞ problem, plus a new formula ψ, let IPSAT_∞ output the (possibly empty or singleton) interval of possible probabilities of the new event ψ given the probabilities r_j of the old events ψ_j, $(j = 1, \ldots, m)$. Evaluate the complexity of IPSAT_∞.

D. Mundici, *Algebraic Probabilistic Consistency*, Trends in Logic 69,
https://doi.org/10.1007/978-3-031-98337-5_11

Index

D. Mundici, *Algebraic Probabilistic Consistency*, Trends in Logic 69,
https://doi.org/10.1007/978-3-031-98337-5

A

B

C